Stephen James O'Meara's
Exploring the Solar System with Binoculars

In this journey of discovery, Stephen James O'Meara shows readers how to observe our Solar System wonders with ease and clarity, using the unaided eye, inexpensive handheld binoculars, or large mounted binoculars.

The book presents a new way to see and appreciate the wonders of the Solar System in detail, including lunar and solar eclipses, sunspots, craters on the Moon, planetary detail, meteors, and comets. It is a unique observing guide for all amateur astronomers proving you don't need big and expensive equipment to enjoy astronomy from your own backyard.

Readers will learn how to find Venus in the daytime, how to observe faint features in bright comets, how to maximize your chances of seeing the most meteors during a shower, how to monitor the changing aspects of the planets and their moons, and much more.

STEPHEN JAMES O'MEARA has spent much of his career on the editorial staff of Sky & Telescope, and is a columnist and contributing editor for Astronomy magazine. He is an award-winning visual observer. His remarkable skills continually reset the standard of quality for other visual observers, and he was the first to sight Halley's Comet on its return in 1985. The International Astronomical Union named asteroid 3637 O'Meara in his honor. Steve is the recipient of the prestigious Lone Stargazer Award (2001) and the Omega Centauri Award (1994) for "his efforts in advancing astronomy through observation, writing, and promotion, and for sharing his love of the sky." He has also been awarded the Caroline Herschel Award for his pre-Voyager visual discovery of the spokes in Saturn's B-ring and for being the first to determine visually the rotation period of Uranus. Steve is also a contract videographer for National Geographic Digital Motion, and a contract photographer for National Geographic Image Collection.

Also by this author:
Stephen James O'Meara's
Observing the Night Sky with Binoculars
A Simple Guide to the Heavens

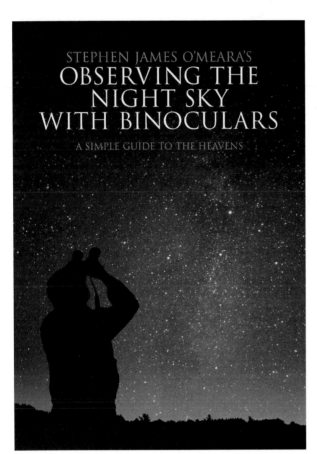

Month by month, star by star, object by object, Stephen James O'Meara takes readers on a celestial journey to many of the most prominent stars and constellations visible from mid-northern latitudes, exploring the brightest and best stars, nebulae, and clusters visible through inexpensive, handheld binoculars.

"The chapters cover all the major binocular deep-sky objects well, but unlike most such guides, the author devotes even more space to star patterns and individual stars. On the whole, I find this refreshing . . . As always, O'Meara's writing is lively, quirky, and infused with his personality." *Sky and Telescope*

"[O'Meara] relates his invaluable experience as an astronomer, revealing the unique stories and secrets each constellation has to offer, bringing them right down to Earth. O'Meara's writing inspires and his passion and enthusiasm for observing leaps off the pages." *Sky at Night Magazine*

"O'Meara's book really did keep me interested from the outset. . . . The mythology and history of the stars and other objects are explained in engaging narrative, and the reader is left feeling as though they have truly learned about what they have seen . . . ideal for those who want to know more about astronomical objects easily seen with binoculars." *Astronomy Now*

"a fine book that should encourage any possessor of simple optical aids to go out and seek for themselves what the night sky has to offer." *The Observatory*

ISBN 978-0-521-72170-7

Stephen James O'Meara's
Exploring the Solar System with Binoculars

A Beginner's Guide to the Sun, Moon, and Planets

CAMBRIDGE
UNIVERSITY PRESS

CAMBRIDGE UNIVERSITY PRESS
Cambridge, New York, Melbourne, Madrid, Cape Town, Singapore,
São Paulo, Delhi, Dubai, Tokyo

Cambridge University Press
The Edinburgh Building, Cambridge CB2 8RU, UK

Published in the United States of America by Cambridge University Press, New York

www.cambridge.org
Information on this title: www.cambridge.org/9780521741286

First published 2010

Printed in the United Kingdom at the University Press, Cambridge

A catalog record for this publication is available from the British Library

Library of Congress Cataloging in Publication data
O'Meara, Stephen James, 1956–
Exploring the solar system with binoculars : a beginner's guide to the sun,
moon, and planets / Stephen James O'Meara.
    p.   cm.
Includes index.
ISBN 978-0-521-74128-6 (pbk.)
1. Solar system – Amateurs' manuals.   2. Solar system – Observers' manuals.
3. Astronomy – Amateurs' manuals.   4. Astronomy – Observers' manuals.
I. Title.
QB501.2.O44   2010
522'.6 – dc22   2010001370

ISBN 978-0-521-74128-6 Paperback

To Donna,
My love orbits around you
To Milky Way, Miranda Piewacket, and Pele,
My spirits in the sky
To Daisy Duke Such a Joy,
My faithful satellite
To William (Bill) Albrechet (1917–2009),
He loved the night

Contents

Preface

In her 1912 book *The Ways of the Planets* (Harper and Brothers; New York), Martha Evans Martin says that to know the planets is to know ourselves, because the Earth is a planet. This was still true when I was young. Astronomers did not know much about the planets. Venus was thought by some to have humid rainforests capable of sustaining life. Mars still had an air of mystery about it that had some scientists clinging to the hope that its soil contained bacterial life – though more and more astronomers were turning their backs on what they believed was a world as dead as the Moon. And who knew about the distant outer worlds, what mysteries they held?

Today, that picture has changed magnificently. It would be more appropriate now to say that to know our Solar System is to know ourselves. Armadas of spacecraft have explored all the major planets of our Solar System, and we now know their surfaces. We know of towering volcanoes and vast canyons on Mars, erupting volcanoes on Jupiter's moon Io. We've seen new ring systems around other planets and have altered the definition of a planet. We no longer look to the Earth as the only abode of life. And we no longer look to our Solar System as the only system with life. New planetary systems are being discovered around other stars at a rapid pace.

Comets, we now know, are creators and destroyers: They may have given our planet water and seeded it with the building blocks of life. We have also witnessed the frightening effects of comets crashing into Jupiter. Likewise, it's possible that meteorites, blasted from the surface of Mars during its early history, could have transported microbes to the surface of the Earth. They may have also led to the extinction of the dinosaurs; and killer asteroids and comets remain a threat to humanity.

We have also grown accustomed to humans as explorers. Only today, instead of conquering mountaintops, we look out into space for our next great challenge beyond the Moon. The Solar System is our hometown, and we enjoy learning about the wonders of our celestial neighborhood.

The Solar System is more than our home, it is a part of human history. The movements of the planets have played with human emotions and beliefs. Their positions against the starry backdrop have led us into battle, or augured apocalyptic events. Eclipses burned into the eyes and souls of our ancient ancestors who looked upon them with abject fear and wonder. Similarly, great comets have swept across the skies like evangelical swords, and meteors have rained down from the sky like celestial tears.

Of all the things in the universe, Solar System objects were, and still are, the most accessible to anyone beginning in astronomy. All these wonders we can still appreciate today with nothing more than our unaided eyes and a pair of binoculars. And, yes, while we cannot see incredible details, say on Mars or Jupiter, we do have our minds. That is why I've seeded this book with some NASA or other spacecraft images of these worldly wonders, because binocular observing is half reality, half imagination. I want you to appreciate the wonders that you see, not only with your eyes, but also with your mind and heart.

I'll never forget my first views through binoculars, when a friend showed me the four moons of Jupiter. I was a young Galileo seeing new worlds for the first time. These binoculars could also zoom to $20\times$; so I was shocked also to see the rings of Saturn (or at least the "ellipse" of the rings) – something that I thought was only possible through a telescope. And how could I have imagined that the most distant gas-giant worlds, billions of miles distant, could be seen as small stars in handheld binoculars. Later, I saw the phases of Venus . . . and then, a comet!

From comet tails to meteor trains, from the phases of Venus to the genesis of sunspots, it's all here in this book, which is intended to bring you all that childhood wonder. Its not so much a field guide as it is a companion that will inform you about what's within the limits of your vision – both with your unaided eyes and through binoculars. It is to be used as a companion to my book *Observing the Night Sky with Binoculars* (Cambridge University Press; Cambridge, 2008).

The book is as strongly dedicated to naked-eye observing as it is to binocular observing, and the detail and training I provide in the book will also help you when, or if, you decide to graduate to a telescope. Naked-eye observing goes hand-in-hand with binocular observing. Can you see Venus in the daytime with your unaided eyes? How about Jupiter? Mars? Binoculars, you will find, will be your trusted friend in helping you to confirm the limits of your naked-eye vision.

For the observations made in this book, I used three types of binoculars: handheld Bresser 7×50s (Bresser.com; also available at Walmart®), handheld Meade 10×50s (Meade.com), and tripod-mounted Orion 25×100s (telescope.com). The two numbers on the binoculars refer to the magnification (the first number) and the diameter of each front objective lens in millimeters (the second number). Thus, 7×50

binoculars have an aperture of 50 millimeters and will magnify an object (like the Moon or planets) seven times. The greater the magnification, the larger an object appears in your field of view. The larger the aperture, the brighter the object will appear. Binoculars 10 × 50 and smaller can be used comfortably without a tripod. Anything larger requires a tripod. Most astronomy magazines support websites with lots of information helpful to first-time buyers. Two useful articles are Richard Talcott's "Using binoculars" (http://www.astronomy.com/asy/default.aspx?c=a&id=2225) and binocular guru Phil Harrington's "Binoculars under $100," which appears in the April 2005 issue of *Astronomy* magazine. Harrington also has a very helpful website (http://www.philharrington.net/). Larger binoculars are especially effective when looking at the Sun through safe solar filters (**never look at the Sun without proper protection; doing so could cause permanent eye damage or blindness**) and the glorious Moon with its myriad detail. Large binoculars also allow you to see faint comets and dim features in their heads and tails; they also reach fainter magnitudes, allowing you to see fainter asteroids and get better views of the colors of the more distant planets. But I have used 10 × 50 binoculars for most of my astronomical career and have found them a trusted friend.

As with any hobby, it takes time to excel. In astronomy, the more you look, the more you will see. So take the time to look and learn. This book is set up in a simple and logical fashion. I begin by looking at the Sun, the nearest star to Earth and the brightest celestial object in the sky. I then help you to explore the wonders of the second brightest object in the sky: the Moon. The Moon is much more than a floating rock; it's the most detailed Solar System body you'll see through binoculars and is reminiscent of the many other moons inhabiting our solar neighborhood. To see the Moon is to see worlds beyond up close and personal. I help you to identify nearly 200 individual features, including its craters, seas, mountains, and more.

In Chapter 3, I investigate the fantastic drama of solar and lunar eclipses. I delve into the histories of these events, and include detailed observations of some little-understood phenomena (there is just so much to see!).

Chapter 4 takes you on an informative tour of the planets, working our way out from the inferior worlds of Mercury and Venus, to blood-red Mars, then on to the tremendous gas-giant worlds (Jupiter, Saturn, Uranus, and Neptune); I also help you to find a bright dwarf planet − a king among the Solar System's minor planets in the Main Asteroid Belt.

Chapter 5 covers the beautiful and exotic comets, reviewing some of the incredible influences they have had on humanity and describing in detail how best to observe them; I also provide descriptions of some of the greatest comets visible to the unaided eyes and binoculars over the last century. The last chapter, introduces you to meteors and meteor showers. I provide ample information on the best showers of the year and how to observe them. I also give details of their histories and the types of displays you can expect from them. Thanks to Gareth Williams, director of the Minor Planets Center at the Harvard-Smithsonian Center for Astrophysics, I've included an appendix of more than 100 bright asteroids to search, especially for those with large binoculars.

Finally, I would like to take a deep bow to Jay Pasachoff (Field Memorial Professor of Astronomy at Williams College) for reviewing the material on the Sun; Fred Espenak (Mr. Eclipse, who recently retired from the Goddard Spaceflight Center) reviewed the chapter on eclipses. Renowned planet observer and author William Sheehan donated his eye to the planets chapter. Dan Green (Director of the Central Bureau for Astronomical Telegrams) lent his excellent assistance on the comets chapter. Meteor expert Peter Jenniskens (Senior Research Scientist at the Carl Sagan Center of the SETI Institute) gave his valuable critique of the meteor section. And Gareth Williams reviewed the section on the minor planets. I would also like to thank Eric Kopit of Orion Telescopes and Scott Roberts of Explore Scientific for their assistance with the binoculars used in this project, my editor Vince Higgs at Cambridge University Press for believing in the book, and Zoë Lewin for her careful eye and sensitive copyediting. I take full responsibility, however, for any slips of the tongue. Finally, I cannot tell you how thankful I am to my wife, Donna, for her patience during the writing of this work.

Stephen James O'Meara
Volcano, Hawaii
April, 2009

CHAPTER 1

The Sun: angel of light

The Sun, with all those planets revolving around it and dependent on it, can still ripen a bunch of grapes as if it had nothing else in the universe to do.

– Galileo Galilei

WARNING!

* Failure to use proper methods when observing the Sun may result in permanent eye damage (retinal burns or thermal injury), which could cause or lead to blindness.
* Do *not* look at the Sun with your unaided eyes for any prolonged period of time.
* Do *not* use sunglasses, smoked glass, CDs, DVDs and CD-ROMs, film negatives, or polarizing filters, to look at the Sun; they are not safe for the purpose!
* Do *not* look at the Sun directly through binoculars without them being covered with specifically designed filters – filters that cover the objective lenses at the front end of the binoculars and let only about one part in 100,000 through, reflecting the rest.
* *Never* leave unattended mounted binoculars (without filters) pointed at the Sun, since even an instant's viewing could be devastating to someone's eyesight.
* Do *not* use solar filters at the eyepiece end of your binoculars. The focused light can burn through, or crack, the filters.
* Purchase solar filters only from reputable dealers.
* If in doubt about a filter's safety . . . *do not use it!* Do contact your local planetarium or astronomy club for more information.

When the ancients finished their nightly musings under the stars, and the Sun brought the light of day, the answer to at least two astronomical questions (what is a star, and what purpose does it serve?) was burning brightly before them. The Sun is the nearest star to the Earth. It is the center of our Solar System and our true Angel of Light.

The Sun is with us every day and will be with us long into the future. Earth cannot escape its gravitational embrace. We are are destined to circle our star year after year, lifetime after lifetime. The Sun affects our moods, warms our bones, and burns our skin. Its light influences the way we see the world and how artists express their feelings. Without the Sun, the sky would not be blue, breezes would not blow, and rains would not fall. All usable energy on Earth – including oil and coal – is

directly, or indirectly, manufactured by the Sun. Plants convert sunlight into food, which enables them to stay healthy and grow. All animals, including humans, need plants, air, and water to survive. Simply put, the Sun is the most important star in the sky; without it, life on Earth would not exist.

Little wonder then that Sun worship has prevailed throughout recorded history. The first Egyptian pyramid, the Step Pyramid at Saqqara, was dedicated to the Sun; its shape may reflect the step-like appearance of the setting or rising Sun during certain mirage conditions. Some of the largest neolithic and pre-Columbian structures in the world – including the Great Pyramid of Giza, Stonehenge, and Machu Picchu – have intimate alignments with the rising and setting Sun, whose position shifts seasonally

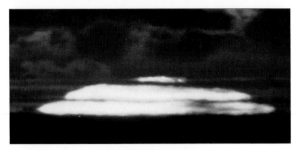

from from upper left) the Step Pyramid, the rising Sun distorted by mirages, Hathor, and the ankh.

The most ancient Mayan deity identified in archaeological records is the Sun God. Many early peoples revered the Sun as the father of all gods or the mother of light and life, such as Australia's aboriginal Sun goddess Yhi. The pharaohs of Egypt believed they were descendants of the Sun god Ra. In ancient Greece, the Sun gods Helios, and later Apollo, fathered all other Greek gods. Likewise, chiefs of Sonabait, who once ruled the Timor region of Indonesia, regarded themselves as the "children of the Sun." And in Japan, the "Great Divinity Illuminating Heaven" was Amaterasu, from whom the imperial Japanese family claims descent. One would be hard pressed to find an ancient culture that did not recognize the Sun as a deity or source of omnipotent power.

The solar powerhouse

along the east and west horizons like a slowly swinging pendulum.

Ancient peoples recognized not only the importance of the Sun's position along the horizon but also its divine power as the creator and nurturer of all living things. Researchers have found depictions of Sun gods and goddesses on nearly every continent. Among the oldest are reliefs of various Egyptian deities carrying the solar disk atop their heads, such as Hathor, the supreme mother of pharaohs. Hathor is often depicted carrying the ankh, which, like the Sun, symbolizes eternal life or resurrection; the ankh's figure may, in fact, depict the Sun cresting the horizon. The images above show (counterclockwise

Meet the Sun	
Magnitude:	−27
Spectral type:	G2 (yellow dwarf)
Age:	~4.6 billion years
Diameter:	1,400,000 km
	(870,000 miles)
Surface temperature:	5,785 K (10,000 °F)
Rotation period (equator):	24 days 6 hours
Inclination of axis:	7° 15′
Mean distance (from Earth):	150,000,000 km
	(93,000,000 miles)

The brightest and closest star in our sky, the Sun (from the Latin, *Sol*) is an enormous glowing sphere of gas at an average distance of 150 million km (93 million miles), or 1 astronomical unit (AU).[1] With a diameter of about 1.4 million km (870,000 miles), the Sun is so large that you could line, side by side, 109 Earths across its equator. If Earth were the size of a baseball, the Sun would be a sphere about 60 meters (200 feet) smaller than the 365-meter-wide (1,197-foot-wide) London Millennium Dome, and about 46 meters (150 feet) larger than the 256-meter-wide (840-foot-wide) Georgia Superdome!

[1] The Earth's orbit around the Sun is not a perfect circle, but an ellipse. Earth's distance varies from about 147 million km (91.3 million miles) around January 3, to about 152 million km (94.4 million miles) when farthest away around July 7.

Open the Sun like a chest, and you could fit 1,300,000 Earths inside. Actually, the Sun is not a perfect sphere. Scientists using NASA's Ramaty High Energy Spectroscopic Imager (RHESSI) spacecraft have measured the roundness of our star with unprecedented precision, and found that during years of high solar activity, the Sun becomes more oblate, i.e. flattened at the poles. For instance, during its peak activity in 2004, the Sun increased its 1.4-million-km (870,000-mile) equatorial diameter by about 13 km (8 miles). It's a puny anomaly, despite which, the Sun is still the biggest and smoothest object in the Solar System – perfect at the 0.001 percent level.

While the Sun is enormous in size, the gases that comprise it – hydrogen and helium (99.9%) and some heavier elements (0.1%) – are so light that a teaspoonful of its matter would be, on average, about one-fourth as light as a teaspoonful of Earth's, or about the density of water.

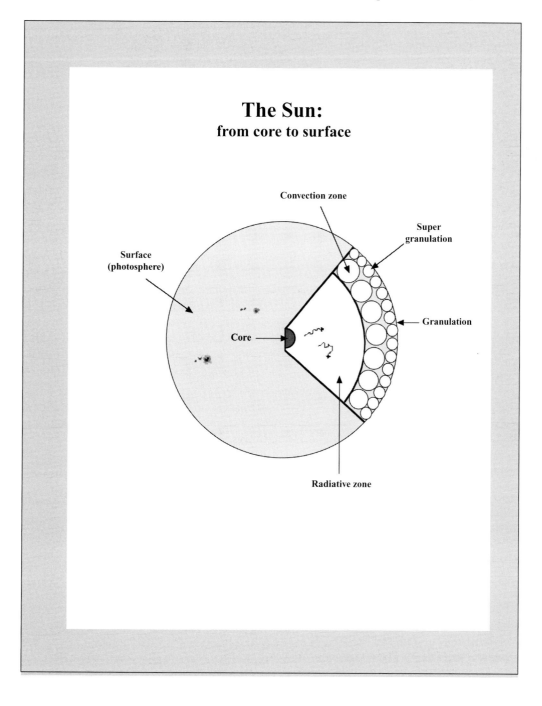

The Sun:
from core to surface

Convection zone

Super
granulation

Surface
(photosphere)

Granulation

Core

Radiative zone

The gases radiate at extremely high temperatures. At the surface, the Sun's heat measures 5,785 K (10,000 °F), or 4.5 times as hot as molten lava. Place the Earth on the "surface" of the Sun and it will vaporize straight away.

The Sun's surface temperature is refreshingly cool compared to the 15.6 million degrees Celsius (28 million degrees Fahrenheit) temperature found at the Sun's core. Here gases are packed so tightly (thousands of tons per square inch) that the nuclei of hydrogen atoms fuse (combine) into heavier helium atoms. Through this process of "hydrogen burning," some 600 million tons of hydrogen are converted into helium every second, a power equivalent to 100 billion hydrogen bombs exploding every second.

The energy radiating outward from the Sun's core is so hot that we'd need X-ray eyes to see it! This energy doesn't just erupt to the surface and spew out into space. It can spend several hundred thousand years inside the Sun, in a region called the "radiative zone," where it is continually absorbed and re-emitted, until it ultimately breaks free and rises to to a region known as a convection zone, where convection is a sort of boiling. As these heated gases approach the surface, they cool to surface temperature and emit visible light. (Most of the light actually comes from electrons attaching and detaching from neutral hydrogen atoms on the solar surface.) They also start to sink until heat from the upper layer of the radiative zone makes them buoyant once again. We see this surface action as granulation, which is described in more detail on page 8. Gases in the Sun's convective zone and surface, then, churn turbulently like boiling water.

While the Sun radiates into space more than half a million tons of energy each second from its surface, only a fraction of that amount reaches the Earth. Still, the solar energy we receive each minute is roughly equal to the amount of electrical energy we artificially generate on Earth each year.

WARNING!

Never look at the Sun directly without proper eye protection. Failure to use proper methods when observing the Sun may result in permanent eye damage (retinal burns) or blindness. Do not look at the Sun directly through binoculars without their being covered with specifically designed filters – filters that cover the objective lenses at the front end of the binoculars and let only about one part in 100,000 through, reflecting the rest. See page 5 for more details.

As viewed through proper and safe solar filters, the Sun actually has three levels of atmosphere visible to the naked eye and binoculars. The *photosphere* (lowest level of the three) is the Sun's visible surface, which we can see on any clear day. Its fantastic brilliance overpowers the weak emission emanating from the other two levels. (The photosphere is described in great detail beginning on page 7.)

The *chromosphere* (middle level) is a spiky 10,000-km-wide (1,200 mile-wide) veneer of gas (less than 1 percent of the Sun's diameter) lying just above the photosphere. Its name, derived from the Greek word *chromos* (color), means "sphere of color." Nineteenth-century astronomers gave it this name because it appears as a thin layer of intense red light around the Moon's black silhouette at the beginning and the end of the total phase of a total solar eclipse. The color comes from the chromosphere's hydrogen gas, the strongest emission of which is hydrogen-alpha in the red part of the Sun's spectrum.[2]

The chromosphere has a temperature of about 5,600 °C (10,000 °F) near its base and 50,000 °C (90,000 °F) at its top. The structure varies from the fine to the majestic. The subtlest features are small jets of gas, called spicules, that bristle up from the Sun's limb like fine neck hair on a cold day. In the nineteenth century, they were sometimes referred to as "burning prairies." Actually, they're transparent magnetic pipes (hundreds of thousands of them at any time) filled with plasma moving at speeds of 50,000 km (31,000 miles) per hour. The features are short-lived, lasting only a few minutes, squirting up and falling down like fairy fountains.

The chromosphere's most majestic features are its *prominences*. These prodigious eruptions of dense gas can lift off from a section of the Sun's limb (i.e. entire circumference) in a variety of forms – from fiery tongues to fantastic hedgerows. These plasma clouds can be held suspended by the Sun's magnetic field for hours or weeks; some seem to appear in concert with violent eruptions of solar flares (see page 18). Many prominences rise into the Sun's outer atmosphere, the ethereal *corona*, whose temperature suddenly rises to nearly 1.7 million degrees Celcius (3 million degrees Fahrenheit). While the photosphere is visible to us every clear day, naked-eye and binocular observers can see the chromosphere and corona only when the Moon covers the photosphere during a total or annular solar eclipse. (Again, I must stress that the details in the chromosphere and corona cannot be seen through binoculars unless you're observing a total or annular solar eclipse. Failure to use proper methods when observing the Sun may result in permanent eye damage, retinal burns, or blindness.) The appearance of prominences and the corona will be discussed in greater detail in Chapter 3.

Despite all the Sun's glory and grandeur compared to earthly standards, the Sun is just an average star. We have found stars much larger than the Sun and many much

---

[2] If you ever graduate to a telescope (and I encourage you to do so one day), you can supplement your solar viewing with a hydrogen-alpha filter, which will reveal the Sun's chromosphere and open a door to a new and ever-changing world of solar wonder. A world where spicules, plages, flares, and prominences suddenly burn forth in crimson splendor. For hydrogen-alpha viewing, consider Coronado's Personal Solar Telescope (PST), which costs around $500 to $600.

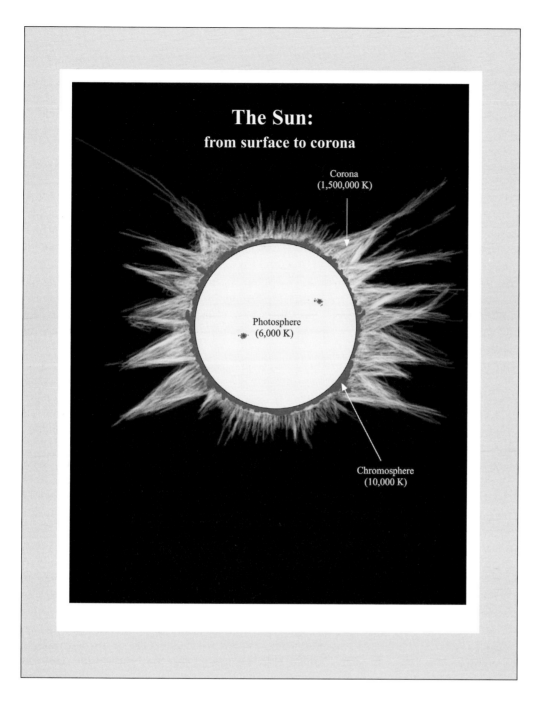

The Sun:
from surface to corona

Corona
(1,500,000 K)

Photosphere
(6,000 K)

Chromosphere
(10,000 K)

smaller; we have also found stars much hotter than the Sun and others much colder. Considering the hundreds of billions of stars in our galaxy alone – and there are trillions upon trillions of galaxies, each with billions upon billions of stars, in the Universe – our Sun has no outstanding qualities... except for one: if it weren't for the size of this unassuming star and its distance from Earth, you would not be around to read these words. The Sun, say Leon Golub and Jay Pasachoff in their 2001 book *Nearest Star: The Surprising Science of Our Sun* (Harvard University Press, Cambridge, Massachusetts; London, England) is our "Goldilocks," meaning that it creates the conditions "just right" for life to flourish on Earth. Is it surprising then we still worship the Sun today in our own private ways?

Like anything that burns energy, the Sun's lifetime is limited. Our star has been consuming its nuclear fuel for nearly five billion years. But it does so very conservatively; the Sun is expected to burn for another five billion years before exhausting its hydrogen supply and will begin to die. Today, as anyone on a cloudless day can see, the Sun's light is still very intense; intense enough that anyone foolish enough to stare directly at it for any length of time without proper eye protection risks damaging (or losing) his or her eyesight!

How to observe the Sun safely

The Sun offers safety-conscious observers highly detailed views of the only star resolvable from Earth. Keeping track of the Sun's ever-changing features can keep one pleasantly occupied for a lifetime. But before I explore the visual splendor of the Sun with you, let's look at two

> **WARNING!**
>
> * Failure to use proper methods when observing the Sun may result in permanent eye damage (retinal burns or thermal injury), which could cause or lead to blindness.
> * Do *not* look at the Sun with your unaided eyes for any prolonged period of time.
> * Do *not* use sunglasses, smoked glass, CDs, DVDs and CD-ROMs, film negatives, or polarizing filters, to look at the Sun; they are not safe for the purpose!
> * Do *not* look at the Sun directly through binoculars without them being covered with specifically designed filters – filters that cover the objective lenses at the front end of the binoculars and let only about one part in 100,000 through, reflecting the rest.
> * *Never* leave unattended mounted binoculars (without filters) pointed at the Sun, since even an instant's viewing could be devastating to someone's eyesight.
> * Do *not* use solar filters at the eyepiece end of your binoculars. The focused light can burn through, or crack, the filters.
> * Purchase solar filters only from reputable dealers.
> * If in doubt about a filter's safety . . . *do not use it!* Do contact your local planetarium or astronomy club for more information.

that you can hold the binoculars with both hands. Of course, if the binoculars can be mounted on a tripod, by all means do so!

> **WARNING!**
>
> *Never* leave unattended mounted binoculars pointed at the Sun, especially in a public setting; someone may become curious, take a peek, and burn his or her eyes.

The further you hold the card from the eyepiece, the larger the Sun's image. With my 7 × 50 and 10 × 50 binoculars, I like to view a disk that is about 5 to 8 cm (2 to 3 inches) across, which requires placing the card some 50 to 80 cm (20 to 30 inches) away from the eyepiece. When I hold the card about 50 cm (20 inches) away from the eyepiece of 25 × 100 binoculars, I get a beautiful image of the Sun some 10 cm (4 inches) across. Experiment until you find the distance that gives you the best view with your binoculars.

You can increase the apparent contrast between the projected image and its surroundings by using a front shield. I made one out of a piece of cardboard with two holes large enough to slip objective mounts through. Wearing sunglasses to look at the projected image will help cut down on the glare reflecting off the white card to your eyes; or use a light gray card.

Try to keep the observations brief enough so that the Sun's heat is not concentrated for long periods of time on the optical system. I generally project an image for a minute or two, then turn the binoculars away from the Sun before repeating the observation a few minutes later.

simple and inexpensive ways you can view it in complete safety.

Projection

The safest and simplest way to view the Sun is to project its image through one lens of your binoculars onto a large white, or light-gray, card or piece of cardboard. No filters are required for this method because you *never* look through the binoculars. Start by capping one of the binoculars' front objective lenses and turning your back to the Sun. Holding the binoculars in one hand, point the uncovered objective lens toward the Sun behind you. With your free hand, place the card about a foot from the binocular's eyepiece. To achieve proper alignment, use the shadow of the binoculars as a guide. Simply move the binoculars around until the Sun's image appears on the card within the binocular's shadow. Focus on the Sun's limb until it looks sharp.

Once you master this technique, repeat the procedure while either bracing the binoculars against something sturdy (like a table, chair, or wall), or propping up and supporting the paper at the appropriate angle (the Sun's projected image should appear circular not elliptical) so

Welder's glass

Proper solar filters are designed to reflect or absorb a specific amount of ultraviolet, visible, and infrared energy – namely one part in 100,000 of light transmitted – for safe solar viewing. One of the most common and inexpensive filters in the Sun observer's arsenal is a shade number 14 welder's glass, which is widely available for a few dollars

from welder supply shops; you can find one in any local phone directory. The glass comes in a variety of sizes and shapes. All those of shade #14 convey an agreeable green image of the Sun.

A 10-cm-square (4-inch-square) glass allows you to observe the Sun with your unaided eyes; make sure the glass covers both eyes at all times. You can also use the glass to cover the front of your binocular objectives. (If the filter is large enough to cover only one binocular objective, *remember to cap the other objective!*). But make sure the filter is securely mounted, and that it doesn't fall off or blow loose while you are looking through the binoculars.

If you purchase a smaller, rectangular welder's glass, I suggest you mount the filter in a cardboard frame that fully covers the binocular objective. The poor optical quality of the glass yields a soft and slightly distorted binocular image. I have observed the Sun with welder's glass both with unaided eyes and in front of binocular objectives for more than a quarter century and have never hurt my eyes, but I am always especially careful when looking at the Sun. The images above show the welder's glass in a cardboard frame (left) and the proper way to hold it up to one front objective (right), as my wife, Donna demonstrates. Always remember to cap the unused objective!

Mounted filters

Many reputable telescope dealers sell special metal-coated glass or mylar filters in mounted cells that fit snugly over, or screw tightly to, most binocular objectives. Like the shade #14 welder's glass, special, mounted solar filters take out the infrared energy. Note that some other filters – including ordinary photographic filters (even so-called neutral density ones) – don't! Always be sure to use a safe and special solar filter!

Using special, cell-mounted filters allow you to look directly at the Sun in comfort and safety with binocular vision! (Always check the filters for surface damage before using them; for instance, mylar – a very thin plastic film with a coating of aluminum – can be easily punctured or ripped.) I used Orion® full-aperture glass solar filters, which are mounted in aluminum cells. The filter's glass

elements are machine-polished and triple-coated with an advanced nickel–chromium alloy that gives the Sun a soothing amber hue. With no optical distortions, the Sun appears as a pleasingly crisp disk with sharp detail, especially through 25 × 100 binoculars. The photo below shows the 25 × 100 binoculars with the solar filters on; on top of them I piggybacked a pair of 10 × 50 *unfiltered* binoculars for size comparison. Again, make sure that any filters are securely mounted or properly taped on; be careful that they don't come loose.

The photosphere

Photosphere, meaning "sphere of light," is derived from the Greek word, *photos*, for light. The Sun has no solid surface, but we commonly refer to the photosphere as its everyday, visible surface. It is what we see when the Sun rises each morning and sets each afternoon. It is the face we see when the Sun's image is projected onto a card, or when we view it directly through properly filtered binoculars. When we look at the photosphere through our filters, we're seeing light fleeing from the Sun's turbulent surface; like all light and radio waves, it is traveling at a speed of 300,000 km (186,000 miles) per second. This light must travel some 150 million km (93 million miles) before reaching our eyes eight minutes later. Consequently, whenever we look at the Sun's filtered disk, we see it as it was eight minutes in the past.

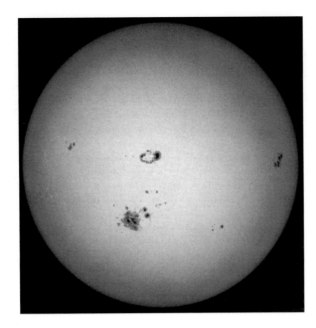

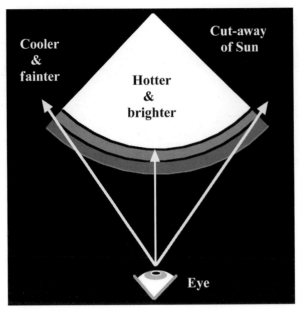

Limb darkening

When you project the Sun's image, or look at its filtered disk through binoculars, survey the intensity of light from the disk's center to its limb. The Sun's surface brightness drops off gradually, then more severely, as you look closer to the limb. The effect is called *limb darkening*, and it is one of the perennial features of the white-light Sun.

Limb darkening is the visible effect on light as it rises from the base of the convection zone to the cooler surface of the Sun's photosphere. As the diagram above (right) shows, when we look directly at the Sun's center, we look a bit into the convection zone — very near its cool top. When we look at the Sun's limb, we do not look at all into the convective zone; instead, we look obliquely through only the cooler upper layers of the photosphere. Since cool gas does not glow as intensely as hot gas, the Sun's limb appears darker than it does at its center; it also appears very slightly redder; in the black-and-white image above (left), limb darkening appears as a darker shade of gray.

Granulation

If you concentrate on the filtered Sun's surface and let your eye flit around the binocular field of view, you may see that the Sun's photosphere is not smooth. It has a fine, granular texture. The detail becomes especially apparent when one is using large filtered binoculars that magnify at least 20× and the atmosphere is very stable.

The image at right is a white-light SOHO satellite image that shows the Sun's mottled surface texture on a totally spotless day in December, 2008. Expect a softer view through your binoculars. You'll find that with time as you observe, the granulation stands out ever more prominently. And once you get sight of it, it's hard not to see it again; remember, the longer you look, the more prominent the granular texture will seem.

Two astute observers — Danish Astronomer Royal Thomas Bugge (1740–1815) and the German-born British observer William Herschel (1738–1822) — appear to have noticed the Sun's granular texture around the same time in 1792. Interest in the phenomenon led many observers to better describe and decipher what they were seeing. One famous interpretation blossomed in 1861, when Scottish engineer/observer James Nasmyth (1808–1890) announced to the Literary and Philosophical Society of Manchester that the Sun's photosphere was dappled with small, elongated features, shaped like willow leaves. He described the leaves as crossing one another in all directions and constantly moving.

This announcement led to a "vehement controversy," George F. Chambers tells us in his 1904 book *The Story of the Solar System* (D. Appleton and Company; New York), which led to the use of other expressions, such as "'rice grains,' 'sea beach,' and 'straw thatching,'" to describe the phenomenon. Chambers thought all these words were too precise to be taken literally; though, "on the whole," he admits, "'rice grains' is not altogether a bad

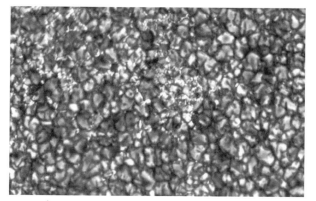 

expression . . . " But it wasn't until 1864, when another Englishman, William Rutter Dawes (1799–1868), coined the term "granulated," which formed the word we still use today.

Inspired by observations by William Herschel (who called them "forrows" which were bordered by "indentations") astronomers began to theorize that the granules were the summits of leaping flames (prominences) all over the Sun's surface – an idea that carried over into the twentieth century! These theories were not far from the truth.

Today we know that solar granules – several million of which are visible at any given time – mark the places where huge (several-hundred-km-wide) bubbles of hot gas are rising up from the base of the hot convection zone and erupting onto the Sun's surface; the granules remain intact for only about 15 minutes or so before they fragment, merge with other granules, or sink, cool, and fade, only to be replaced with new bubbles of hot, rising gas. We see this perpetual activity as a slowly shifting "rice grain" pattern across the entire face of the Sun. So the granules are the convection regions that I discussed earlier.

Look closely with your filtered binoculars and you'll see that each bright luminous patch is separated from its neighbors by less dark *intergranular lanes*, where cooling gases descend. So the surface looks like a bubbling broth of rice pudding seen from afar. Because the Sun is so distant, the individual "grains" can be a challenge to see through small, filtered, handheld binoculars. Granulation is best seen with a direct view through mounted binoculars whose objectives are covered with safe solar filters.

Through mounted 10 × 50 binoculars capped with solar filters, I can "sense" granulation when I slowly move the Sun around in the binocular field of view, or gently tap the tube. Tube tapping is an old observing trick. When you stare directly at an object without shifting your gaze, vision fades (which is why the eye has natural rapid eye movements). Tube tapping provides slight motions, allowing the object in view to sweep back and forth across the eye's photoreceptors, so that continuous (non-fading) images are sent to the brain for processing. (That's why it's easier to see a gnat in flight than when it's resting, say, on a leaf.)

Although I do not see *individual* grains clearly through 10 × 50 binoculars, I do get fleeting impressions of a hyperfine roughness across the entire Sun's face, superimposed on which are some more obvious blemishes, which may be clusters of granulation. The Sun does have larger-scale supergranulation cells which can measure some 16,000 to 32,000 km (10,000 to 20,000 miles) in diameter – but you don't really see those in white light. Actually, the average size of a solar granulation is *about* 1.5 arcseconds across, which is near the limit of 10 × 50 binoculars and often below the steadiness of the air above you.[3]

I get the same impression, though one much more difficult to savor, when I use 10 × 50 binoculars to project the Sun's image onto a card; be sure to make the Sun's image as big as possible and focus the binoculars until the disk appears sharp. Success, however, also requires mounting your binoculars (with a Sun shield) on a tripod, projecting the Sun's image onto a white or gray card, blocking any extraneous light, and moving the card around so that the Sun's sandpaper texture is not confused with the tiny fibers making up the card or paper you're holding.

Fine solar granulation is definitely visible when I look directly through the mounted 25 × 100 binoculars with safe solar filters. The individual grains are fantastically small, looking like heaps of hot and cooling molten flecks. Indeed, I have seen molten rock bubbling in the throat of a volcanic vent and have noticed that it is eerily similar to seeing these roiling clouds of hot gases that rise to the surface of the Sun. The images above show a high-resolution image of the Sun's granulation (left) taken with the Swedish 1-meter (40-inch) Solar Telescope and processed

---

[3] Resolving power is defined as an instrument's ability to see fine detail, like granulation on the Sun. To determine your instrument's resolving power, divide 4.56 by the aperture in inches. The answer, known as Dawes' limit, gives you the resolution in arcseconds ("). One arcsecond is 1/60 of an arcminute ('), or 1/3600 of a degree. (The Sun measures about 30' across.) A pair of 50-mm (2-inch) binoculars can resolve details to 2.3" (aside from limits caused by Earth's turbulent atmosphere), while a pair of 100-mm (4-inch) binoculars can detect features as small as 1.2". But this limit applies to bright points of light seen against a dark background. The eye can resolve finer details through the same instrument used in the daytime! So solar granulation is indeed within the grasp of 50-mm binoculars, and certainly resolvable in 100-mm binoculars.

in a computer, which shows much more detail than you'll see in your binoculars (but knowing what you're actually seeing is part of the imagination side of astronomy). To the right is an image showing the surface of a molten lava lake for comparison. Through 25 × 100 binoculars, the granules can be spied with a direct gaze and without any tube tapping. Looking at these granular features and thinking about the action forming them truly makes the Sun come alive.

Sunspots

As exciting as it is to see solar granulation, nothing on the Sun's surface compares to catching sight of its dark spots. These mysterious clouds of gas can range in size from about 2,400 km (1,500 miles) across to a colossal 48,000 km (30,000 miles) across; the latter size being great enough to swallow four Earths. The image below shows a close-up of a large sunspot group with an artificial silhouette of the Earth for comparison. It also shows a full-disk, white-light SOHO image of the Sun, which represents well the detail one can see through filtered binoculars.

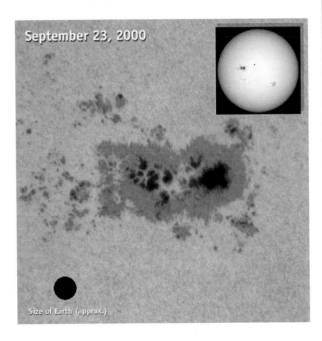

September 23, 2000

Size of Earth (approx.)

the Shang Dynasty (from about 1500–1050 BC) habitually observed the low Sun. According to their interpretation of inscriptions on oracle bones (animal bones and turtle shells inscribed with a primitive form of Chinese characters), they also made the "earliest written records of sunspots." For instance, one inscription says, "There was ri zhi in the western sky, will it bring a disaster?" The authors note that ri means "the Sun," while they interpret zhi, to mean "a black spot" or "black vapor." Using a welder's glass or other safe viewing filter, you too can keep vigil on the Sun and record naked-eye spots.

Below is an illustration of a record of naked-eye sunspots as seen through a #14 welder's glass; it's based on a page from one of my notebooks. The sketches here show the naked-eye Sun in June 1989, near sunspot maximum – a time when sunspot activity peaks in its roughly 11-year cycle. (The cycle is described in more detail on page 14.)

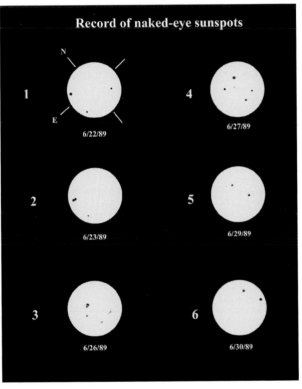

Large sunspots (as well as some smaller ones) are visible to the unaided eyes through a safe solar filter. It's also possible to see them without a filter when the Sun is near setting and the atmosphere is especially contaminated with dust or other pollutants, which can greatly dim the Sun's brightness. Many ancients made their observations of the Sun in this latter way.

In a 1995 *Quarterly Journal of the Royal Astronomical Society* (volume 36, pp. 397–406), Xu Zhen-tao (Purple Mountain Observatory, China), F. R. Stephenson (University of Durham, UK), and Jiang Yao-taio (Nanjing University, China) noted that Chinese skywatchers in

Optical sunspots

The invention of the telescope in 1608, and Galileo's turning a telescope skyward in 1609, brought the heavens into a new light. (These telescopes were no better than a good pair of binoculars today.) With the telescope, the spotted Sun was soon noticed by several European observers at about the same time. Galileo Galilei (1564–1642), was one of them. In a letter dated June 12, 1612, the great Italian observer wrote to his patron Giuliano de'Medici that "celestial discoveries are not yet at an end, it is about

23 months and more since I began to see some dark spots on the sun."[4]

The first known telescopic drawings of sunspots (below) belongs to the English mathematician, Thomas Harriott (1560–1621) – the first Englishman to make a telescope and turn it on the heavens – who logged their appearance on December 8, 1610. Harriott's scientific manuscripts, however, remained out of sight, buried in private archives, until Austrian astronomer Baron von Zach rescued them in 1784 and called attention to Harriott's astronomical observations.

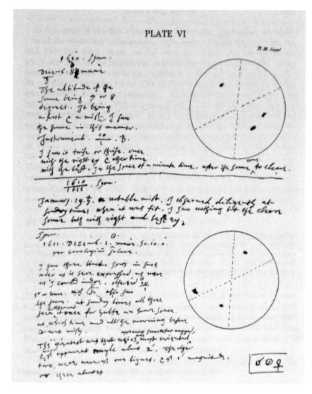

After their telescopic discovery, observers turned their attention to the nature of the spots. Some believed they were aggregations of clouds floating in a solar atmosphere. Others believed them to be small intramercurial worlds seen in silhouette while orbiting the Sun. (Interestingly, the great German astronomer and mathematician Johannes Kepler (1571–1630) made a pre-telescopic observation of a naked-eye sunspot, though he mistook it for a transit of Mercury.) But it was Galileo who astutely placed sunspots on the solar surface. In a letter to Giuliano de'Medici, dated June 23, 1612, Galileo fastidiously documented his thoughts on the origin of sunspots and their behaviors:

[4] Galileo had a long and bitter quarrel with a German Jesuit astronomer, Christopher Scheiner, who (with the assumed name of "Appelles latens post tabulam") claimed that he, not Galileo, first discovered the Sun's spots sometime in March or April, 1611. Actually, the German astronomer Johannes Fabricius (1587–1616) was the first known person to publish a discovery account of the Sun's telescopic spots; the work was dedicated on June 13, 1611, and describes his telescopic discovery of the spots on March 9, 1611. Of course, as I had mentioned earlier, Chinese skywatchers from the Shang Dynasty had habitually recorded sunspots with their unaided eyes when the Sun's light was greatly diminished around the times of the Sun's rising or setting.

[T]hese spots are not only near the sun, but contiguous to its surface, where continually some are being produced and others dissolved . . . They continually change shape . . . Often one divides into 3 or 4, and at other times 2, 3, or more, coalesce into one. They have moreover a regular movement, by which they are all uniformly carried around [by the solar] globe itself, which revolves in a lunar month, with a motion similar to that of the celestial sphere, that is from the west to the east. These spots never occur near the poles of the suns's rotation, but only about the equator, nor are they found further from it than about 28 or 29 degrees, as far toward one pole as another . . .

Sunspots, we know today, are regions on the Sun's surface that appear dark because the gases in them are more than 555 °C (1,000 °F) cooler than their bright surroundings; seen alone against a dark sky at the Sun's distance, a large sunspot would rival the full Moon in brightness as seen from Earth. If you look carefully at a sunspot, you'll see it has a dark central region, called an *umbra* (from the Latin meaning "shadow"), surrounded by a lighter, feathery *penumbra*.

The penumbra's filamentary structure is reminiscent of the pattern formed by iron filings around a bar magnet. Indeed, sunspots are associated with intense magnetic fields; they generally occur in pairs with the magnetic field coming out of one and into the other. The image below shows a sunspot with unprecedented spatial resolution. It was taken in May, 2002, with the Swedish 1-meter (40-inch) Solar Telescope and shows inner structure, a dark core, and finely resolved filaments in sunspot penumbrae. So when you turn your filtered binoculars to the Sun and see the delicate penumbral features, you can better appreciate the magnificence of what you're truly seeing.

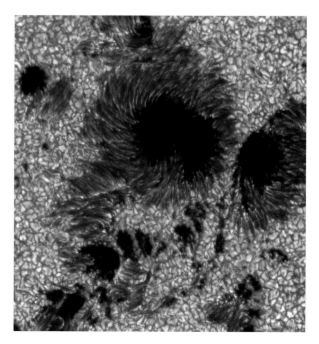

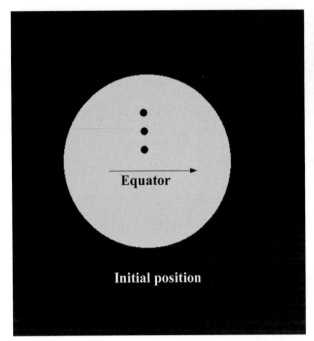

Initial position

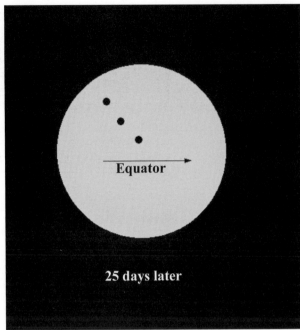

25 days later

Despite being observed for 400 years, sunspots remain largely an enigma. But we are rapidly learning more about them, thanks, in part, to research conducted with instruments aboard the Earth-orbiting Solar and Heliospheric Observatory (SOHO) and other spacecraft. Using SOHO data, Alexander Kosovichev and Junwei Zhao of Stanford University, along with Thomas Duvall of NASA's Goddard Space Flight Center, have discovered that sunspots are surprisingly shallow, lying on top of swirling, planet-sized hurricanes of electrified gas. These storms plunge surface material downward toward the spot's center, pulling magnetic fields together as the material descends. This action creates an intense magnetic "plug" beneath the sunspot that suppresses convection and prevents heat from reaching the surface – which explains why sunspots are cooler and appear darker than their surroundings. As long as the magnetic field under a spot remains strong, the cooling effect will maintain an inflow that makes the structure stable. Eventually, however, the field weakens, the system destabilizes, and the spot decays.

Surface rotation

Unlike a solid planet, like the Earth, whose entire surface rotates at the same rate (~24 hours in a day), the Sun has a gaseous surface that rotates at different speeds – a phenomenon known as *differential rotation*. Gases near the Sun's equator rotate faster than those at the poles. As measured north or south from the Sun's equator, surface material at, say, 10° solar latitude, will rotate faster than surface material at 20° solar latitude, which rotates faster than that at 30° solar latitude, and so on.

Surface material near the Sun's equator takes about 25 days to make a complete turn around the the Sun's imaginary axis of rotation. Surface material close to the pole takes more than 30 days to make the same trip. Recall, though, how much larger the Sun is than the Earth. While

the equatorial material at the Sun's surface takes 25 days to complete one rotation, it is traveling at a speed of about 7,200 km (4,500 miles) per hour! Likewise, then, the speed at which a sunspot travels across the Sun's face, depends on its latitude.

A sunspot first appears on the Sun's eastern limb and sets on its western limb. Looking directly through your filtered binoculars with north up, the rotation of the Sun carries spots from the left (east, as in terrestrial east) side of the disk to right (west). If you project the image with your binoculars, the "up" and "down" directions will remain the same, but "left" and "right" will be reversed when you look down at the projected image (as the top illustration on page 13 shows). Therefore, a sunspot you see on the Sun's left limb on the projected image will be on the right limb when looking directly through filtered binoculars.

Unfortunately, it's difficult for binocular observers to pinpoint exactly where on the Sun's limb its rotational axis lies, making it equally challenging to determine the orientation of the Sun's equator. Consequently, when we see a sunspot through our binoculars, it requires some mental gymnastics to imagine exactly where (north or south of the Sun's equator) sunspots lie.

There are three main reasons for this confusion:

First, the Sun's equator is inclined about 7.5° to the plane of Earth's orbit. So, as the Earth travels around the Sun, we see a substantial seasonal tilt of the Sun's axis due to the ever-changing perspective. Over the course of the year, the Sun makes one complete wobble, like a top, as shown in the bottom diagram on page 13.

Note how, on March 7, we see the Sun's southern pole at maximum tilt toward the Earth, which causes the Sun's equator to appear north of the disk's center. The Sun's northern pole achieves maximum tilt towards the Earth around September 8th. On January 5 and July 7, the Sun's equator cleanly cuts the Sun in two equal halves with both its north and south poles clinging to the limb; only

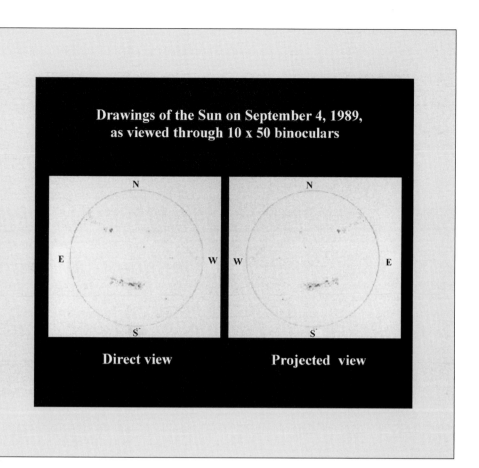

Drawings of the Sun on September 4, 1989, as viewed through 10 x 50 binoculars

Direct view   Projected view

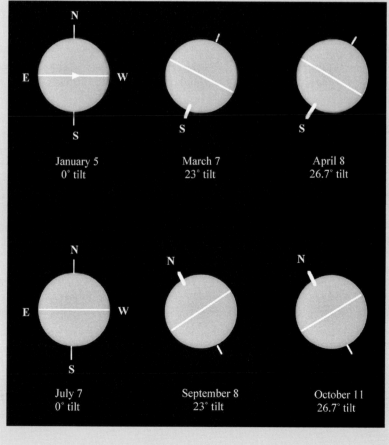

Variations of the positional angle of the Sun's axis during the year

January 5
0° tilt

March 7
23° tilt

April 8
26.7° tilt

July 7
0° tilt

September 8
23° tilt

October 11
26.7° tilt

on these latter two dates do sunspots appear to move across the Sun's disk in a straight line.

Second, as the Earth rotates, and the Sun rises and sets in the sky, the orientation of the Sun's polar axis shifts relative to the horizon, as depicted in the photo illustration below.

Third, how you position your head when you look through your filtered binoculars will also affect disk orientation; a slight tilt of the head slightly to the left or right of vertical will cause the orientation of the Sun's axis to shift in the field of view.

But do not despair. We are not living in an age of ignorance. Technology is on our side, so use it. If a sunspot or sunspot group becomes visible, and you want to know its latitude north or south of the equator, search the Internet for NASA- and ESA-sponsored sites relating to the Sun: One great place to start is sohowww.nascom.nasa.gov/home.html, which provides real-time images from the SOHO spacecraft with proper north–south orientation. Just click on the link to "sunspots" and you'll see the day's "Michelson Doppler Image (MDI) Continuum" image, which shows the Sun in white light; the view approximates what you would see through a good pair of binoculars.

Another great (and highly recommended) place to start is Dr. Tony Phillips' NASA-sponsored site www.spaceweather.com. It offers daily news and information about the Sun–Earth environment; the text is understandable to anyone just beginning in astronomy and valuable to experts as well. Among its wide variety of topics that should interest backyard observers, the site includes a daily white-light SOHO spacecraft image of the Sun, pertinent information about any visible sunspots, and a number of excellent links to other essential sites. Checking either of the sites just mentioned (you could also go to www.solarmonitor.org) before you go out to look at the Sun will help you to prepare for your observations and eliminate confusion.

Spot cycles and migration

Over the years, you'll discover that the Sun's face is not always covered with spots. Rather, the activity waxes and wanes in a roughly 11-year cycle, with the number of sunspots growing from a minimum (no spots) to a maximum (dozens of spots) in about four or five years, then back to a minimum about six or seven years later, on average. The SOHO images below show the Sun at maximum in 2001 (left) and at minimum in 2008–9 (right).

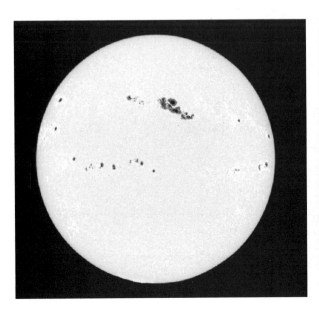

Samuel Heinrich Schwabe (1789–1875), a German pharmacist and amateur astronomer, was the first to announce a roughly ten-year sunspot cycle. In an 1843 *Astronomische Nachrichten*, he explained how, after observing the Sun with a small telescope since 1826 on every clear day, he had noticed that the Sun sported no spots on about half the days of observation in 1833 and 1843. This contrasted sharply with observations he made around 1828, and again around 1838, when he saw spots every day he looked. But it took at least another solar cycle before the world accepted his results.

When a new cycle of solar activity begins, sunspots start forming at "high" latitudes – generally around 35° north or south of the Sun's equator. (Some rogue spots have been recorded at latitudes as high as 70°, but never at the poles.) As the years pass, the area of spot formation migrates closer and closer to the Sun's equator, until the region of spot formation reaches a solar latitude of around 15° and a new spot cycle occurs. At this time, new spots begin forming once again at high latitudes,[5] while spots from the old cycle continue to congregate near the Sun's equator. Eventually the old spots just peter out. For a time, then, it's possible for old and new spots to coexist.

Known as *Spörer's law* – after German astronomer Gustav Spörer, who had described the phenomenon by the 1860s – this gradual equatorward drift of sunspots

[5] Although you cannot see it, when a new sunspot cycle begins, the leading sunspot in each sunspot group in the Sun's northern hemisphere takes on a certain polarity (either negative or positive). It's the same way with sunspots in the Sun's southern hemisphere, only that the common polarity is reversed. When the next sunspot cycle begins, the polarity of the leading members in each sunspot group reverses. So while you visually observe the waxing and waning of sunspots in an approximate 11-year cycle, keep in mind that it's only a part of the 22-year magnetic cycle that governs solar activity.

during a solar cycle was actually first recognized by the English amateur astronomer Richard C. Carrington (1826–1875), who made a careful study of sunspots from November 1853 to March 1861. As early as 1858, Carrington published in the *Monthly Notices of the Royal Astronomical Society* a paper titled "On the Distribution of the Solar Spots in Latitude since the Beginning of the Year 1854; with a Map." In that work, he writes,

> [T]hroughout the two years preceding the minimum of frequency in February 1856, the spots were confined to an equatorial belt, and in no instance passed the limits of 20° of latitude [north] or [south;] and that shortly after this epoch, whether connected with it or not, this equatorial series appears to have become extinct . . . and two new belts of disturbance abruptly commenced, the limits of which in both hemispheres may be set between 20° and 40° . . .

The findings of Carrington and Spörer were beautifully augmented by an intense investigation by British solar astronomer Edward Walter Maunder (1851–1928). On May 8, 1903, Britain's Astronomer Royal communicated a paper that summarized Maunder's research on sunspot number and location spanning the years 1874 to 1902. The results, which were presented in both tabular and graphical form, were based on about a quarter of a million measurements (made in duplicate) of some 5,000 separate sunspot groups as they appeared on about 9,000 photographs.

The following year, however, Maunder revised his approach to the data in the *Monthly Notices of the Royal Astronomical Society*, publishing it in a new graphical form. Now popularly known as Maunder's *butterfly diagram* (below), it shows that when the number and position of the spots are

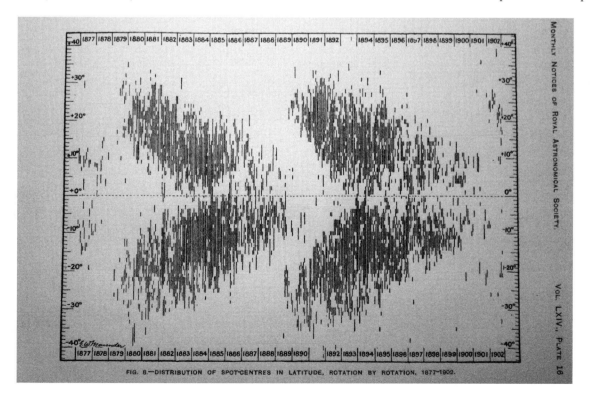

FIG. 8.—DISTRIBUTION OF SPOT-CENTRES IN LATITUDE, ROTATION BY ROTATION, 1877–1902.

plotted against time for both solar hemispheres, the shape of the distributions resembles the wings of a butterfly.

In 1893, Maunder made another significant discovery. While looking back at the historical record of sunspot activity, he noticed that sunspots were largely absent between the years 1645 and 1715. This *Maunder minimum*[6] was a virtual meltdown of the sunspot cycle, which occurred, most curiously, in concert with the middle of the coldest period of the Little Ice Age – an intense episode of climate change that caused global temperatures to drop 2 to 3 °F from today's average.[7] The cooler weather also led to severe hardships, especially in western Europe and parts of North America as crops froze, livestock died, and people perished from famine and disease.

In their 2001 book *Nearest Star* (Harvard University Press; Cambridge, MA), Leon Golub and Jay M. Pasachoff note that Maunder's work was forgotten for decades until John Eddy of the High Altitude Observatory drew attention to it in the 1970s, reminding us that if the sunspot cycle could shut down once, it could do it again. This would imply, the authors note, that the sunspot cycle is superficial rather than fundamental to the Sun. The question is, when will the next Maunder minimum occur . . . if at all?

Golub and Pasachoff also cite research by Sallie Baliunas and others at Mount Wilson Observatory, which establishes that other stars also have solar activity cycles – including some solar-type stars that appear to be in a Maunder minimum phase of very low activity. "Enough observations of this type," express the authors, "could establish how much of the time the Sun is in a protracted minimum of activity, by determining what fraction of the solar-like stars are in such a state at any given time."

Sunspot count

Sunspots belong to what we call the "active Sun." The granular surface of the Sun described on page 8 belongs to the "quiet Sun," when the 11-year cycle of sunspot activity is at a minimum. But the face of the Sun is quite active with spots during the sunspot maximum. As you can see on the image at bottom left on page 14,

[6] Adding to the confusion of solar discoveries, while Spörer's law should be called Carrington's law, the Maunder minimum should actually be called the Spörer minimum, since it was Spörer, not Maunder, who first drew attention to the lack of sunspots between 1645 and 1715. But Spörer also noticed a period of low sunspot activity between 1420 and 1570, which coincides with another period of low global temperatures during the Little Ice Age; that episode of low solar activity is now appropriately known as Spörer's minimum.

[7] While solar activity may have compounded global cooling during the Little Ice Age, it is unlikely the sole cause. Throughout the Little Ice Age, which lasted, arguably, for at least half a millennium, heightened volcanic activity (whose sulfur dioxide gas clouds covered the stratosphere with sulfuric aerosols), and a temporary shutdown of large-scale ocean circulation, may have also been contributing factors. (Man definitely was not a contributing factor to the climate change in this pre-industrial-revolution world.)

sunspots appear not only as singular spots, but they also congregate in groups. The number of sunspots you see on any given day will depend on several factors, including instrument aperture, magnification, personal judgment, atmospheric stability, and experience. Just how many spots are on the Sun?

While there are various, sophisticated modern methods of determining the daily sunspot count, you will do just fine by adopting the simple and efficient method first devised in 1848 by Swiss astronomer Johann Rudolf Wolf (1816–1893), director of Bern Observatory. Wolf, whose solar research was inspired by Schwabe's discovery of the sunspot cycle, devised a simple yet effective way to tally the number of spots on the Sun based on the number of groups and individual spots seen through a small instrument.[8]

Still known today as the *Wolf number* (though many sources refer to it as the *relative sunspot number*), it provides a reasonable and reliable indication of the Sun's overall activity; it is especially suitable for the backyard solar observer who wants to keep a personal tally.

To calculate the Wolf number, use the equation

$$R = 10g + s$$

where R is the Wolf number, g is the number of groups on the Sun, and s is the number of individual spots. If you see four groups of sunspots with 15 spots (see below), for instance, then the Wolf number (R) for that day would be $R = (10 \times 4) + 15$, or $R = 55$. If there is one spot on the Sun, then $R = (10 \times 1) + 1$, or $R = 11$.

[8] Schwabe's discovery of the sunspot cycle also inspired Wolf to investigate all the historical data about sunspots. After tracing the sunspot cycle through the former centuries, Wolf refined the sunspot cycle's period to a mean value of $11\frac{1}{9}$ years, but with large irregularities between 7 and 17 years.

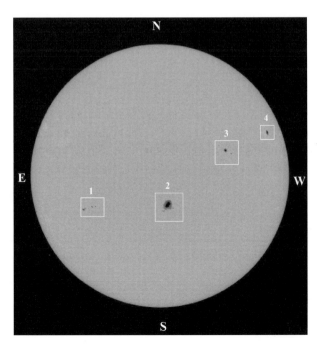

Sunspot groups and classification

As the sunspot cycle progresses, especially as it nears the maximum, you may find it difficult to assign certain spots to specific groups. The conundrum is not new. The earliest such classification was devised in 1901 by the Rev. Aloysius L. Cortie of Stonyhurst College Observatory. Max Waldmeier modified Cortie's scheme in 1938. Known as the *Zurich System* of sunspot classification, Waldmeier grouped sunspots into nine lettered classes according to their physical appearance and extent. This system also took into account whether the spots are *unipolar* (a single spot, or compact cluster of asymmetrical spots), or bipolar (two or more spots forming an elongated cluster of 3° length in heliocentric longitude.). Then in 1966, solar astronomer Patrick McIntosh of the NOAA Space Environment Lab in Boulder, Colorado, revised the Zurich system, and this is the system most used today.

The McIntosh system retains the Zurich system in modified form, while extending it to 60 distinct types of groups "without demanding much of the solar observer." It's a three-letter system that defines groups on the basis of whether penumbra is present, how penumbra is distributed, and by the length of the group. In contrast to the previous systems, a judgment of complexity is not required. The first parameter of the McIntosh system places sunspot groups into seven modified Zurich classes, as described below.

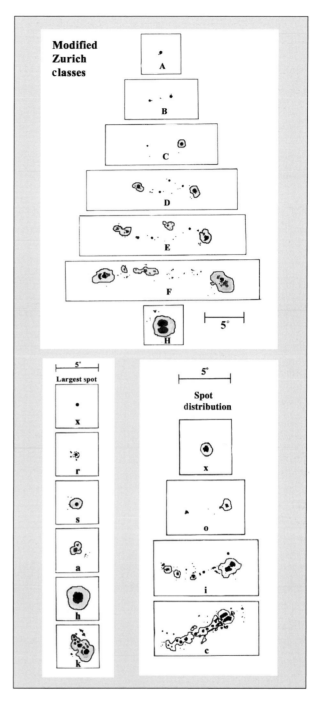

McIntosh's modified Zurich class

Class A: Unipolar group with no penumbra.

Class B: Bipolar group without penumbra on any spots.

Class C: Bipolar group with penumbra on one end of the group, in most cases surrounding the largest of the leader umbrae.

Class D: Bipolar group with penumbra on spots at both ends of the group, and with length ≤ 10° solar longitude.

Class E: Bipolar group with penumbra on spots at both ends of the group, and with length as: 10° < length ≤ 15°.

Class F: Bipolar group with penumbra on spots at both ends of the group, and with length > 15°.

Class H: Unipolar group with penumbra. The principal spot is usually the leader spot remaining from a pre-existing bipolar group.

The system then uses a second letter to describe the type of the largest spot in the group, based on the type of penumbra, size of penumbra, and symmetry of penumbra and umbrae within that penumbra. These descriptors specify size, maturity, stability, and complexity in simple terms, as described below.

Penumbra classification of largest spot

x: No penumbra (group is class A or B).

r: Rudimentary penumbra partially surrounds the largest spot. The penumbra is incomplete, granular rather than filamentary, brighter than mature penumbra and extends as little three arc seconds from the spot umbra.

s: Small, symmetric (like Zurich class J). Largest spot has a mature, dark, filamentary penumbra of circular or elliptical shape with little regularity to the border. There is either a single umbra, or a compact cluster of umbrae, mimicking the symmetry of the penumbra. The north–south diameter across the penumbra is ≤2.5°.

a: Small, asymmetric. Penumbra of the largest spot is irregular in outline and the multiple umbrae within it are separated. North–south diameter of penumbra ≤2.5°.

h: Large, symmetric (like Zurich class H). Same structure as type 's', but north–south diameter of penumbra ≥2.5°.

k: Large, asymmetric. Same structure as type 'a', but north–south diameter ≥2.5.

The final parameter of the three-letter system ranks the relative spottiness in the interior of a sunspot group.

Sunspot distribution

x: Individual spot.

o: Open. Few, if any, spots between leader and follower. Interior spots of very small size. Class E and F groups of *open* category are equivalent to Zurich class G.

i: Intermediate. Numerous spots lie between the leading and following portions of the group, but none of them possess a mature penumbra.

c: Compact. The area between the leading and following ends of the spot group is populated with many strong spots, with at least one interior spot possessing a mature penumbra. The extreme case of compact distribution has the entire spot group enveloped in one continuous penumbral area.

Solar flares (white light)

Solar flares are spontaneous and catastrophic eruptions of energy from the Sun's surface visible to the eye. These violent events – which can heat material on the Sun's surface to many millions of degrees and release as much energy as a billion megatons of TNT – are caused by sudden changes in the Sun's magnetic field. They occur in areas where the magnetic field lines are concentrated, such as near sunspot groups, and may last a few minutes to a few hours.

In 2006, James Drake and his colleagues from the University of Maryland proposed that this sudden change occurs when the concentrated magnetic field lines are squeezed past a critical point. Drake says to imagine a magnetic field line shaped like a 0. If you twist the 0 into an 8, then keep twisting, you'll eventually split the 8 into two 0-shaped fields. While this splitting process normally takes months to occur, it can (for reasons not yet fully understood) accelerate dramatically and create a flare.

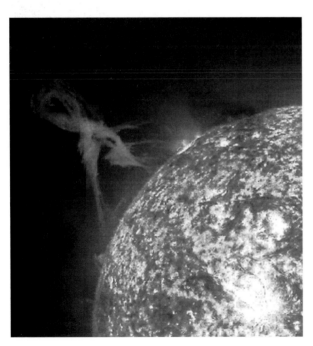

Solar flares are accompanied by the release of gas, electrons, visible light, ultraviolet light, and X-rays. What the binocular user should be interested in are the occurrences of *white-light flares*. This brightening event occurs in the photosphere (so it is visible to the eye), but also in the lower chromosphere.

Flares are one of the powerhouses driving magnetic disturbances on Earth; the other powerhouse is a coronal mass ejection – a huge plasma bubble erupted from the Sun's corona (sometimes in concert with solar-flare events) that travel into interplanetary space. These latter events are the most powerful solar explosion, releasing, over the course of several hours, up to 100 billion kilograms of multimillion-degree plasma at speeds ranging from 10 to 2,000 km per second.

While coronal mass ejections can only be observed from space, strong white-light flares can be seen in large, mounted binoculars with safe solar filters. McIntosh has found that the high energetic flares in hydrogen-alpha and X-ray light have a high probability with types **E** and **F** in his modified Zurich **s** and **k** classes in penumbral shape, and **c** class in distribution. McIntosh also notes that complex sunspot groups move faster in solar longitude before flare activity commences. This may indicate flares are triggered by the fast drift of individual sunspots within the area's complex magnetic configuration. Several models of flare production exist, though there is no consensus.

Curious aspects of sunspots and their surroundings

The Wilson effect

Sunspots first appear on the Sun's eastern limb, rotate onto the disk then disappear on the west edge of the Sun. If you happen to see through your binoculars a particularly large sunspot with a well-defined umbra and penumbra, watch this progression carefully; you may notice a phenomenon first observed in 1769, by Scottish astronomer Alexander Wilson of Glasgow.

In November of that year, Wilson noticed that a great, round spot on the Sun, equally encompassed on all sides by penumbra, had changed its appearance in a most curious fashion as it neared the Sun's western limb. The penumbra nearest to the Sun's center, he noticed, appeared to thin as the spot approached the limb, where it vanished altogether. In contrast, the penumbra nearest the Sun's limb remained practically unaltered as the spot moved. The reverse happened when the spot reappeared at the Sun's eastern limb about two weeks later.

Wilson explained the phenomenon by suggesting that sunspot umbrae were conical cavities that funneled to different depths below the Sun's surface. As rotation carries a sunspot toward the solar limb, the angle at which we see the cavity changes, causing its depression to be revealed. Wilson's original drawing shows how the giant sunspot altered its presentation after it reappeared on the Sun's eastern limb in December 1769. The drawing below, from George F. Chambers' classic 1904 *The Story of the Solar System* (D. Appleton and Company; New York), shows how a giant spot typically changes its appearance as it nears the Sun's western limb.

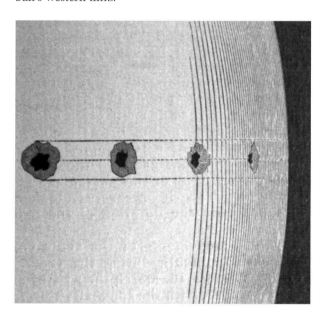

After careful observation, Wilson deduced that the depth of a spot often amounted to about one-third of the Earth's radius, calling it an "excavation in the luminous matter of the sun." Interestingly, while the Wilson effect had been considered an optical illusion for many years throughout the twentieth century – one due to different absorptions of light at different depths within the Sun's atmosphere – we have now come full circle.

In 2006, Hugh Hudson, a research physicist at the University of California at Berkeley, used data from NASA's Ramaty High Energy Spectroscopic Imager (RHESSI) spacecraft data to validate Wilson's interpretation of the effect. Hudson's analysis proved, by direct observation, that the Wilson effect really is due to a physical displacement. Wilson depressions are around 1,000 km deep, while shallow ones are about 200 km deep. Hudson notes that it's not possible to see to the bottom of a deep depression unless the spot is very large, since the curvature of the solar diameter is much larger than that of a typical spot.

Faculae

When a sunspot is rotating onto the Sun's face, you may see it through filtered binoculars surrounded by bright filaments and irregular patches of light. Galileo was well aware of these bright clouds, and, in 1613, called them *faculae*, which is Latin for "little torches." Galileo used faculae, as well as the Sun's spots, to determine the Sun's rotation.

Centuries of solar observations have led astronomers to recognize several common characteristics of faculae and their relationships with sunspots. First the presence of these bright clouds often announces a sunspot's birth, appearing as a compact glow a few hours before a spot's appearance. Faculae remain compact for a few days then begin to spread out. Their greatly irregular and branching forms make them look like some exotic form of solar coral. Faculae attain their greatest extent when a spot is about to vanish and can linger for months after a spot has disintegrated. After that, they start to break up and disappear. The clouds are most pronounced when seen against the darker solar limb, where contrast is enhanced. The SOHO image below shows faculae as they might appear through your binoculars (left) and as they do in calcium light.

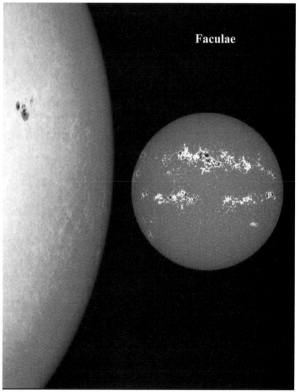

Faculae

Like sunspots, faculae are associated with strong magnetic fields. Unlike sunspots, they represent photospheric hot spots. Faculae form as small magnetic bundles (much smaller than those that form sunspots) in the valleys between solar granulation. The strong magnetic field at their locations reduces the density of the gas, making it nearly transparent; we can look more deeply into the

Sun where the gases are hotter, so faculae appear brighter than the surrounding surface. The yin-yang association of sunspots and faculae is curious, since sunspots decrease solar radiance while faculae increase it. While faculae form in the photosphere, they can at times extend into the chromosphere, where they form plages. Formally known as flocculi, plages are bright spots in the solar chromosphere, usually found near sunspots. They can be viewed only in hydrogen-alpha light and should not be confused with white-light faculae.

Sunspot color

It's often stated that sunspots have a black umbra and a gray penumbra. But this is simply a matter of contrast. If the Moon's limb passes near a sunspot during a solar eclipse, or if you happen to see one of the inferior planets (Mercury or Venus) pass near a sunspot during a transit over the Sun's disk, the sunspot's umbra then looks gray, and the penumbra even paler, when compared to the inky black disk of the Moon or planet's silhouette.

Light bridges

Sometimes, large binocular viewers may see the dark umbrae of sunspots divided by luminous bridges. It's common for novice observers to confuse them with solar white-light flares. Although light bridges appear brighter than the surrounding photosphere, they are not. Both are of the same magnitude. Light bridges appear brighter due to simultaneous contrast effects, which create an illusion of enhanced intensity.

Sunspot genesis

One of the most fascinating things to see on the Sun is the genesis of a sunspot, or, better yet, one trying to form.

For a few days in early October 2008, two solar pores in the Sun's southern hemisphere ebbed and flowed from

view in a matter of minutes (see drawings below). Also known as proto-sunspots (those with no penumbral features), pores are tiny structures that correspond to the first stage in the development of sunspots. I watched the spectacle through 25 × 100 binoculars; the pores were too small to appreciate in 10 × 50s. They appeared, at first, as two smoky gray patches, swimming in a sea of solar granulation. Each pore was surrounded by faculae and joined by a diffuse penumbral line. About a minute later the bridge vanished and the faculae got weaker, but the two pores became noticeably darker. Another minute later, the pores and their associated faculae faded from view and did not return.

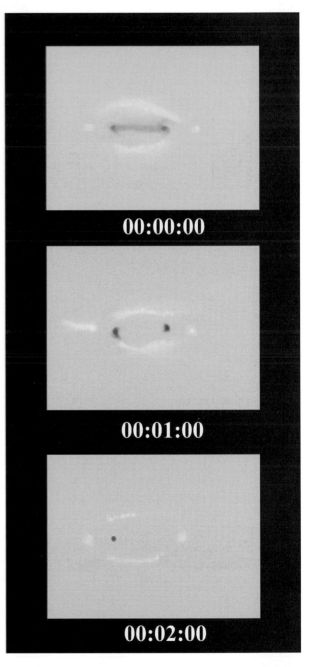

Rapid changes to proto-sunspots and regular sunspots is perhaps the most remarkable thing to see on the white-light Sun. But it requires frequent and patient study of the solar surface over the course of years.

On October 11, 2008, I had the fortune to witness some rapid changes in a small sunspot pair (and numerous pores) through 25 × 100 binoculars (they were a bear to see in 10 × 50s). The drawings below reveal how the two spots and their associated pores struggled to survive.

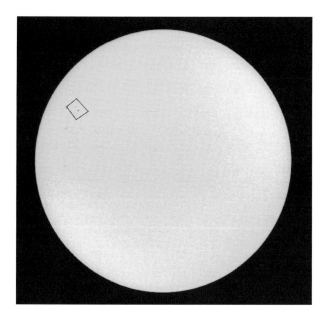

When I first observed them, the two main spots had clear penumbral collars. Each was also accompanied by a dark pore nearby. But as the minutes passed, the pores slowly winked in and out of view. Sometimes, as one pore faded, another would grow at a different location. Several minutes into the observing session, the penumbras surrounding the main spots (as well two pores) disappeared immediately before my eyes – as if by a magician's sleight of hand.

The most incredible event, however, was when one pore (labeled *a* in the drawing) began *violently* pulsing in and out of view in a matter of seconds! The expulsions looked like puffs of smoke chugging out of a volcanic vent. (Having seen and heard such volcanic action, my mind effortlessly recalled those breathy chugging sounds as I viewed the pulsing spot.) At one point, the dark pore was replaced by a bead of intense white light before the pore instantly dissolved.

Shortly thereafter, the umbral portion of the leading spot split into three distinct fragments, making it appear elliptical in a north–south line. But this odd apparition was also short-lived, as the structures morphed back into a single spot that continued to battle for its life. The drama was so intense that I had to wonder if my eyes were being fooled by an unstable atmosphere. So I turned my 5-inch

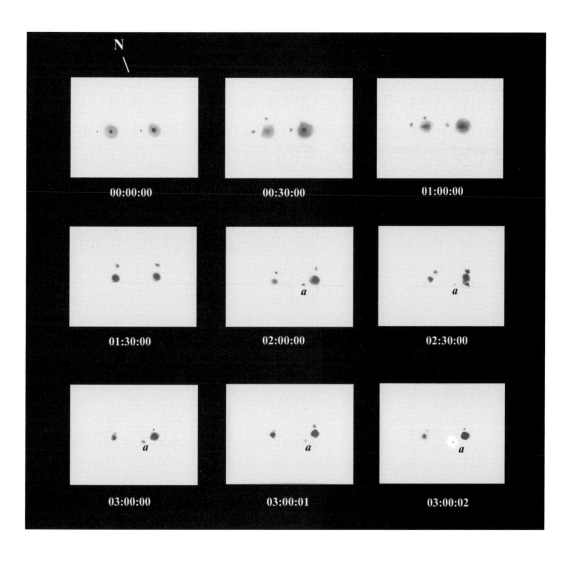

f/5 Tele Vue refractor with a safe solar filter onto the spots and viewed them at high power. The drama continued. It was not Earth's atmosphere undergoing changes — it was the Sun's! So few things in the heavens undergo changes so rapidly before one's eyes as does the Sun's surface, which is why solar observing is such a great pastime.

Not only can sunspots and pores change second by second, minute by minute, hour by hour, but also day by day. The day after the two spots mentioned above formed, only one of them, and one pore, survived — at least that's how the situation appeared during the few minutes I got to see them before clouds rolled in. The point is, the Sun's a book that's constantly turning a new page. Make a drawing of the Sun every clear day for a week, and you can see not only the spots change but also the Sun rotate. In other words, you can experience what it must have been like for Galileo to see the spots behave in their steadfast, yet mysterious, ways.

In amateur astronomy, and especially when making observations of features like sunspots and other temporal solar events, keep good records and be sure to include the date and time of the observation, as well as any other pertinent notes, such as the quality of the seeing or cloud cover. And again, always be concerned about eye safety; always be aware of the condition and security of your solar filters, especially when they are mounted on your binoculars.

CHAPTER 2

The Moon: the lovely dead

surface was crystalline in nature, and that it reflected the features of Earth's surface, which were visible as muted dark and bright patches across the Moon's face. The Moon was then, in a sense, the keeper of Earth's soul.

Actually, we see the Moon because it reflects the Sun's rays. Earthlight (called *earthshine*) can also illuminate the night side of the Moon. But we cannot see a reflection of the Earth in its surface. The Moon has also reflected our superstitions. In Timothy Harley's 1970 book, *Moon Lore* (Charles Tuttle Co.; Rutland, Vermont), the author tells us it has been an abode for both birthing and receiving human spirits. It has also been a penal colony for sinners, who must spend an eternity behind the bars of moonbeams.

It's as dark as charcoal, yet it lights up our night. Its surface has been burned by molten fire, blasted by asteroids, and uplifted into lofty peaks. It has no air, no caressing breezes, no liquid water, no grasses to grab with your toes. Yet humans have looked upon it with unrivaled wonder since the dawn of consciousness. This harsh but alluring world is the Moon, the goddess of our night, Earth's nearest natural satellite.

While romantics may view the Moon as a symbol of purity and unattainable desires, through the cold eyes of science, it is a dead world, one as bleak as a tomb, and as cold and lifeless as a corpse. Yet it's our Moon, and we seem to love the way its light flirts with the landscape and our passions.

What poet hasn't gazed upon the Moon and tried to verbalize its luminous spirit? Who hasn't stared into its spotted face, as if looking into a mirror that could reflect the lamp of his or her soul? Our ancestors of old did believe that mirrors possessed the power to show the soul. They also believed that the Moon's

All cultures have imagined figures on the Moon, the most famous being the Man in the Moon – a delightful arrangement of dark and bright surface features that appears to the unaided eye as a grinning face. Harley tells

us that Plutarch, in his treatise, "Of the Face appearing in the roundle of the Moone," cites the poet Agesinax as saying of the orb:

> All round about environed
> With fire she is illumined:
> And in the middes there doth appeere,
> Like to some boy, a visage cleere;
> Whose eies to us doe seem in view,
> Of colour grayish more than blew:
> The browes and forehead tender seeme,
> The cheeks all reddish one would deeme.

But the lunar surface features have also been imagined as a Woman, a Hare, and a Toad, among other fanciful creatures and personages. For instance, Harley refers to a Talmudic tradition that Jacob is in the Moon. And in a more sinister portrayal, the Moon's bone-white face with its dark eyes and toothy grin, is a dead ringer for a skull. Indeed, when the Moon is low – especially around the times of rising or setting when pollutants can redden its face, and especially in the fall, when the ecliptic is low to the horizon after sunset – seeing a face in an orange full Moon rising, says television's *Stargazer*, Jack Horkheimer, may have prompted people to carve faces on pumpkins around the time of Halloween. The images above show the normal Moon (left) and the woman in the Moon (right) seen in profile.

Shape shifting (Moon phases and more)

Unlike the planets, which have Roman and Greek names, moon is a Germanic word related to the Latin *mensis* (month) – a reflection of the Moon's age-old importance as a time piece. The ancients noticed that the Moon is not always full. As it moved eastward among the stars, they saw it changing phases, completing a full cycle every $29\frac{1}{2}$ days. Thus the first calendars were lunar ones, comprised of lunar months.

When the Moon was low in the west after sunset, our ancestors saw it shining as a thin crescent – as the old Moon (the dusky section illuminated by earthshine) in the new Moon's arms (the crescent illuminated by sunlight) (see the bottom right photo on page 23). As the days progressed, and the Moon marched ever eastward, the crescent grew ever larger, or waxed, as if growing by the accumulation of candle wax. When the waxing Moon was due south at the time of sunset, they saw its shape become half full, a phase now called first quarter. The Moon then became more egg-shaped (a gibbous moon), until it rose full and bright in the east just as the Sun was setting in the west. After full Moon, the phases reversed themselves (waned) until, after last quarter Moon, a gradually narrowing crescent faded in the dawn before sunrise.

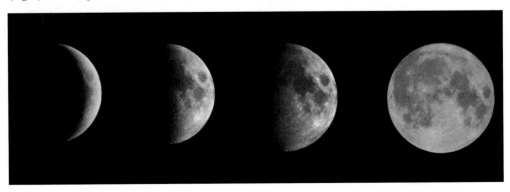

If breaking a mirror is bad luck . . . then the Moon has been cursed from birth. The mirror of the Moon is constantly being shattered, only to be pieced together once more, before it shatters again. It is an endless cycle mimicking that of life and death. The perpetual drama is the celestial equivalent of Sisyphus's task of rolling a rock up a hill, only to have it fall back down again, over and over again. We even see the drama unfold in the nursery rhyme of Jack and Jill.

> Jack and Jill went up the hill
> To fetch a pail of water;
> Jack fell down and broke his crown,
> And Jill came tumbling after.

The rhyme is based on an old Scandinavian Moon myth that plays on the changing phases of the Moon and the belief that they can affect rainfall. When Jack and Jill (spots on the Moon) go up the hill, the Moon is waxing; when they fall down, it is waning. And in the 1995 book, *The Jewish Religion* (Oxford University Press; Oxford), Louis Jacob notes that the waning and waxing of the Moon has been associated with the fate of Israel, which is compared in the Talmud to the Moon.

Even the supernatural idea of shape-shifting may be related to the Moon's phases. Ancient pagans believed that rolling naked in the grass on the eve of a new Moon was one way to become a werewolf – new Moon being, of course, the dark gateway that separated the end of one life and the beginning of another. In medieval times, it was the influence of the full Moon that caused those bitten by a werewolf to become a werewolf. This too makes sense, because what is the old Moon in the new Moon's arms but the beginning of a transformation that culminates during full Moon? To break the curse, one needed to sever a limb from the beast – symbolic perhaps of breaking "the crown," so the beast comes tumbling back down to human form.

There's nothing supernatural about the Moon's phases. We see them because the Moon is a sphere – 3,475 km (2,160 miles) in diameter seen at an average distance of 384,400 km (240,000 miles) – that orbits the Earth.[1] Half of the Moon is always reflecting sunlight, but the amount we see from Earth depends on the geometry. When the Moon is between the Earth and Sun (new Moon), the side of the Moon facing us is in complete darkness. When the Moon is on the side of the Earth opposite the Sun (full Moon), its face is fully lit. At first and last quarter, we see the Moon half lit, and so on. Alas, despite the famous and poetic title of the Pink Floyd CD, there really is no "dark" side of the Moon . . . only a near and far side.

The apparent size of Moon's face will also vary depending on where the Moon is in its orbit. The Moon does not round the Earth. It travels in a slightly elliptical orbit, so the

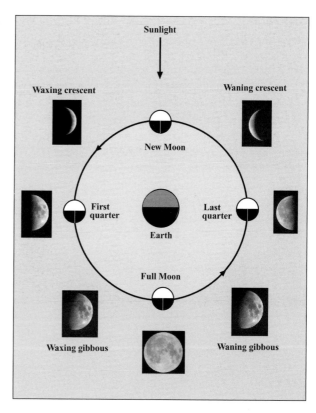

Moon is sometimes closer to Earth than at other times. When farthest away, at *apogee*, the Moon is 405,500 km (251,970 miles) distant. When closest, at *perigee*, it is 363,300 km (225,740 miles) distant. The difference is apparent to the naked eye, being similar to the difference between the size of a quarter and that of a nickel seen at the same distance.

This is one reason why the Moon appears abnormally larger in the sky than at other times. The size of the Moon can also be magnified in the mind's eye when seen close to the horizon near familiar objects of known size – an optical effect, known as the *Moon illusion*.

---

[1] The Moon takes $27\frac{1}{3}$ days to orbit the Earth. But it takes $29\frac{1}{2}$ days to complete one cycle of phases, because the Earth is also moving around the Sun, so it takes the Moon phases a couple of days to "catch up."

There are other delicate changes to the Moon's face that a perceptive observer can catch. For instance, it's common knowledge that the Moon keeps the same face to us as it orbits Earth. That's because the Moon rotates once on its axis in the same amount of time that it takes to complete one revolution around the Earth. But owing to lunar *librations* (slight nodding and rocking motions of the Moon as viewed from Earth), we can see, over time, nearly 60 percent of the lunar surface.

The nodding (*latitude libration*) occurs because the Moon's axis is tilted about $6\frac{1}{2}°$ to the plane of its orbit; so we get alternating glimpses of its poles. The east–west rocking motion (*longitude libration*) occurs because the Moon's speed varies in its elliptical orbit, but its rotation rate doesn't; the net result is that we see the Moon appear to rock gently back and forth as it orbits us. There are other minor librations too, caused by the Moon's placement in the sky, but the longitude and latitude librations are the most significant – especially to a binocular observer, who, over time, can see limb features expand and foreshorten as the Earth and Moon dance.

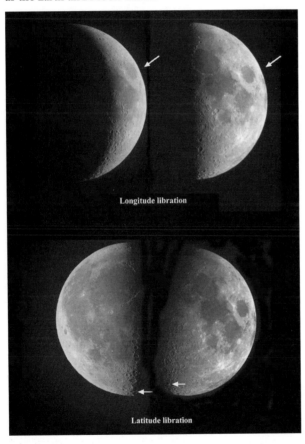

Longitude libration

Latitude libration

The lunar surface

No matter how illustrious we paint the Moon in our imaginations, it is a visually rigid world with features bold and unwavering. Its features change in appearance only when the line between sunlight and shadow (the terminator) sweeps over them as the Moon phases change. Despite its allure, the worst time to see fine details on the Moon is when it is fully illuminated. At full Moon,

shadows all but vanish. Remember, we are looking "down" on the Moon as if seeing the lunar landscape from a distant spaceship passing over it. And just as you see more detail in a terrestrial landscape looking down from a jet at sunrise or sunset, when long shadows play across the terrain, than you do with a Sun high overhead, the Moon's surface is most "alive" when shadows are present.

Meet the Moon

Apparent brightness:	−12.7 (mean)
Diameter:	3,475 km
	(2,160 miles)
Apparent diameter (arcseconds):	1,896 (mean)
Mean relative density (water = 1):	3.35
Weight; if 100 lb on Earth:	17 lb
Mean distance (from Earth):	384,000 km
	(240,000 miles)
Mean orbital velocity:	1 km (30 miles)
	per second
Surface temperature (daytime):	127 °C (+260 °F)
Period of revolution:	27.32 Earth days
Length of day:	29.53 Earth days
Inclination to Earth's orbit:	5.1°
Axial tilt:	6.7°
Number of moons:	0
Rings:	No

If you raise your binoculars to the Moon you'll immediately see two outstanding surface features: bright *highlands* and dark *maria* (Latin for seas). Lunar highlands (or *terrae*, a Latin word meaning lands) are rough-and-tumble patches of highly cratered terrain, the structures of which are best seen in binoculars when the lunar terminator passes over them. These are the vestiges of primordial lunar crust.

The top photograph on page 27 shows the full Moon with the names of the most prominent dark maria (*mare* is the singular form of the Latin word) and the bright lunar highlands. Note the extremely romantic (and dreary) names for the maria, which reflect the early thinking that these areas were truly seas, oceans, bays, marshes, and lakes.

Astronomers believe that our Moon most likely formed when a Mars-sized body slammed into our young Earth some 30 to 50 million years after its formation 4.5 billion years ago. This fantastic collision blasted countless chunks of rock and hot vapor into orbit, which eventually coalesced to form a molten Moon. After cooling, the Moon's primitive surface was pelted by space debris (meteorites, asteroids, and comets), before molten rock welling up from cracks in the lunar surface began to flood and erase some of the impact scars.

Millions of craters still exist on the Moon. They range in size from some 2,000 km (1,240 miles) across to those less than a millimeter in diameter. But only the largest and most magnificent craters are visible in hand-held binoculars. Still, to see the marred face of another world – scars created during episodes of natural violence – is like looking upon a wounded animal in the wilderness,

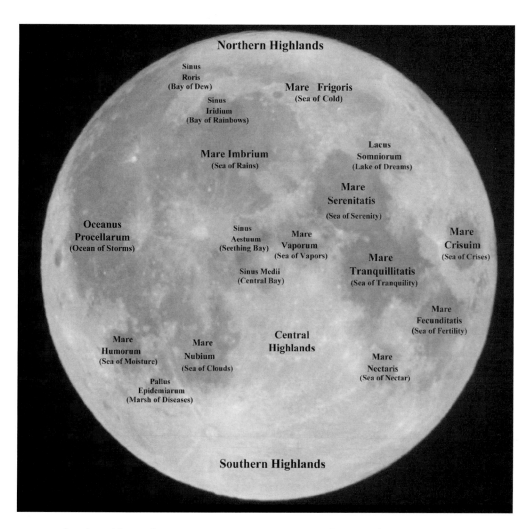

Northern Highlands

Sinus
Roris
(Bay of Dew)

Mare Frigoris
(Sea of Cold)

Sinus
Iridium
(Bay of Rainbows)

Lacus
Somniorum
(Lake of Dreams)

Mare Imbrium
(Sea of Rains)

Mare
Serenitatis
(Sea of Serenity)

Oceanus
Procellarum
(Ocean of Storms)

Sinus
Aestuum
(Seething Bay)

Mare
Vaporum
(Sea of Vapors)

Mare
Crisuim
(Sea of Crises)

Mare
Tranquillitatis
(Sea of Tranquility)

Sinus Medii
(Central Bay)

Mare
Fecunditatis
(Sea of Fertility)

Central
Highlands

Mare
Humorum
(Sea of Moisture)

Mare
Nubium
(Sea of Clouds)

Mare
Nectaris
(Sea of Nectar)

Pallus
Epidemiarum
(Marsh of Diseases)

Southern Highlands

open and vulnerable to future attacks. When we turn our binoculars to our nearest neighbor in space we see not only the ancient memories of episodic extraterrestrial bombardment (the lunar craters, seen below) but also a face scorched by molten fire.

fluid lava flows welled up from the young Moon's interior after being bombarded episodically by space debris. The lava flooded the impact basins and other lowland regions before cooling into solid rock some three to four billion years ago. Over the millennia, these ancient lava beds have become mantled with pulverized rock, called regolith, which can measure 20 m (65 feet) thick in places. Some impact-basin rims were lofty enough to rise above the molten madness, appearing now as tall and rugged lunar *mountains* (photo below).

If the lunar highlands are the "yin" of the Moon, then its maria are the "yang." These soft and dark "flatlands" are vast seas of frozen basalt. As I mentioned, these once

The Moon's mountains are not like Earth's Himalayas or Alps (thrust high into the sky by tectonic forces), but the broken battlements of impact craters. We can see these craggy hills and mountains – some of which rise to magnificent heights [some 8,000 meters (26,000 feet)] above the bleak lunar plains – through our binoculars as ragged ramparts, crumpled masses, and frayed semicircular arcs rimming the Moon's maria. While the mountain ranges can be seen in handheld binoculars (even some to the unaided eyes), they appear most dramatic when sunlight rises and sets on them – when the peaks cast their longest shadows. Impact craters can also be found in the maria, but not to the extent in which we see them in the older highlands. (The Apollo 17 image at bottom right on page 27 shows Montes Carpatus and Mare Imbrium with the grand crater Copernicus at top.)

Now let's look at the Moon, phase by phase, and examine some of the most dramatic features visible in binoculars as sunlight and shadow play with them. It's important to remember that sunlight is constantly on the move (as is the Moon's rocking and nodding motions) so the view that you see here will unlikely be the same when you look with your binoculars, so be flexible in your thinking. The more detailed descriptions are especially for those using large, mounted binoculars that magnify 20× or greater, and especially at times when shadows in and around the features described bring out their subtle highlights. The Moon photos in this chapter follow the modern astronautic convention in which east is to the right and west is to the left – opposite those directions in the sky.

Waxing crescent (1)

A. Mare Crisium (Sea of Crises) is the most conspicuous mare on the face of the the young, waxing Moon. Measuring 418 km (260 miles) across, its dark and flat lava-filled floor is rimmed by tall mountains. It can, at certain librations, appear more oval due to foreshortening effects. The bay to the sea's right is Promontorium Agarum. Two prominent craters grace the sea's western side: Peirce (to the upper left of the label A in the photograph) and Picard (to the label's lower left). Look closely to the left of Peirce and you can see the bright, sinuous, wrinkle ridge Orsum Oppei. The lava flows of Mare Crisium date to 4.5 to 3.8 billion years, during what's known as the Pre-Imbrian period – the date of one of the

Moon's five large impact events that modified the lunar surface.

B. Mare Fecunditatis (Sea of Fertility) is just emerging from darkness in the photo, which reveals part of the sea's 909-km- (565-mile-) wide floor, which appears lightly wrinkled. The lavas of this sea date to the Late Imbrian epoch (3.8 billion–3.2 billion years ago).

1. de la Rue is a lunar crater remnant, or several merged craters, forming a walled plain of unknown depth 134 km (83 miles) across. Named for British astronomer Warren de la Rue (1815–1889).

2. Endymion is a large crater 123 km (76 miles) across, with a smooth, dark floor, which is rimmed by crater walls that rise 2.6 km (1.6 miles) high. Named for the Greek mythological character (a beautiful youth), who spent much of his life in perpetual sleep.

3. Atlas is named for the Greek mythological Titan who supported the heavens. Atlas is a prominent crater 87 km (54 miles) in diameter, with a depth of 2 km (1.2 miles). What can't be seen in this image is its near equal crater, **Hercules**, just to its west; Atlas is part of a delightful crater pair when seen under the right illumination.

4. Franklin is a pretty little circular crater 56 km (35 miles) across and 2.7 km (1.7 miles) deep. It's named for the American inventor Benjamin Franklin (1706–1790) and nicely paired with the smaller, 40-km- (25-mile-) wide crater **Cepheus** one crater diameter to the northwest.

5. Messala is another interesting walled plain with a relatively smooth floor. It extends 125 km (78 miles) in diameter and has a depth of 1.1 km (0.7 miles). At certain lunations, it can appear quite foreshortened due to libration effects. The crater is named for Jewish astronomer Ma-Sa-Allah (unkown–c. 815).

6. Geminus is a circular crater 86 km (53 miles) in diameter and 5.4 km (3.3 miles) deep. It appears slightly oval due to foreshortening effects near the Moon's limb, and its floor has a central ridge. The crater's rim is surrounded by a spray of bright ejecta. Geminus marks the top of a north–south trending chain of three prominent craters leading to Mare Crisium (A), which appears to dangle from the chain like an earring. The crater is named for the Greek astronomer Geminus (unkown–c. 70 BC).

7. Burckhardt is a complex crater of slightly irregular form 57 km (35 miles) across and 4.8 km (3 miles) deep. It lies between Geminus to the north and Cleomedes to its south. The crater's floor has a central rise. It's named for German astronomer Johann Karl Burckhardt (1773–1825).

8. Cleomedes is a large and beautiful crater with a flat floor, small central peak, and a smaller impact feature (Tralles) on its northwest rim. It lies just south of Burckhardt and just north of Mare Crisium (A) and measures 126 km (78 miles) across and 2.7 (1.7 miles) deep. It's named for the Greek astronomer, Cleomedes (unknown–c. 50 BC).

9. Macrobius is a nice circular impact feature with a central mountain complex just beyond the northeastern shores of Mare Crisium (A). It measures 64 km (40 miles) across and is 3.9 km (2.4 miles) deep. It's named for the Roman writer Ambrosius Aurelius Theodosius Macrobius (unkown–c. 410).

10. Taruntius is a prominent crater on the northwestern shores of Mare Fecunditatis (B). It measures 56 km (35 miles) across and is 1 km (0.6 miles) deep. Its floor is fractured and has a low central peak. It's named for the Roman philosopher Lucius Firmanus (unkown–86 BC).

11. Langrenus is a beautiful walled plain and very most prominent impact feature on the eastern side of Mare Fecunditatis (B). It's a circular depression 132 km (82 miles) across and 2.7 km (1.7 miles) deep, but can, at times, appear foreshortened. Its floor is covered with highly reflective lunar dust and craggy peaks, so it shines brightly under a high Sun. Its central peak rises to a majestic 3 km (1.9 miles). The crater is named for the Belgian selenographer and engineer Michel Florent van Langren (c. 1600–1675).

12. Goclenius is a shallow egg-shaped crater on the southwest side of Mare Fecunditatis. It measures 54 × 72 km (33.5 × 45 miles) across and 1.4 km (0.9 miles) deep, and its floor has a low central rise. Through large binoculars, and under the right lighting conditions, you may see the 240-km- (149-mile-) long rille Rima Goclenius extending to the northwest from the crater's wall at the 11 o'clock position (this feature is in shadow in the photo and, thus, cannot be seen). The crater is named for Rudolf Gockel (1572–1621), a German physicist, doctor, and mathematician.

13. Colombo is a circular impact crater 76 km (47 miles) across and 2.4 km (1.5 miles) deep. It's northern rim is disturbed by smaller Colombo A, which has a nice central rise. The crater's floor has been partly covered with lava. The ghost crater to its lower right, with a thin wall and lava-covered floor is Cook. Colombo is named for the Spanish explorer Christopher Columbus (c. 1451–1506).

14. Vendelinus is an an irregularly shaped impact scar on the eastern edge of Mare Fecunditatis (B). It is also old, with its 147-km- (91-mile-) wide dark floor showing multiple wounds from younger impactors, hills and clefts. The crater is 2.6 km (1.6 miles) deep and it walls are well worn and broken in places, making it a fascinating site, especially as it emerges from shadow. The crater is named for Belgian astronomer Godefroid Wendelin (1580–1667).

15. Santbech is an interesting walled plain 64 km (40 miles) across and 4.5 km (2.8 miles) deep. The ancient crater dates 4.5 to 3.8 billion years ago during the Pre-Imbrian period – the same as Mare Crisium (A) – and it has a rough-and-tumble floor with ample detail. It is named for Daniel Santbech (unknown –1561), a Dutch mathematician and astronomer.

16. Wrottesley is a roughly circular crater – 57 km (35 miles) across and 2.3 km (1.4 miles) deep – with a flat floor and twin central peaks. It lies on the northwest fringe of the crater Petavius (below). It is named for British astronomer Baron John Wrottesley (1798–1867).

17. Petavius is a beautiful crater spanning 100 km (62 miles) and 3.4 km (2.1 miles) deep. Under a low Sun, it rules the surrounding crater kingdom, with a complex central crater complex. It is very difficult to see under a high Sun. But under a low Sun, a fantastic rille (Rima Petavius) can be seen running from the crater's peak complex to the southwest crater wall. The crater is named for Denis Pettau (1583–1652), a French chronologist and astronomer.

18. Snellius is a nearly circular impact scar southwest of Petavius. It measures 83 km (52 miles) across and 3.5 km (2.2 miles) deep and has a high central peak. It is a near twin of Stevinus (see below) to its southwest, and the two make an attractive pair under a low Sun. The crater is named for Willebrod van Roijen Snell (1580–1626), a Dutch mathematician, astronomer, and optician.

19. Reichenbach is a nonchalant but noticeable crater 71 km (44 miles) across and 4 km (2.5 miles) deep. It is named for German optician Georg von Reichenbach (1772–1826).

20. Stevinus is a near twin of Snellius to its northeast. Its floor is brighter than that of Snellius's and it has steep inner walls and a fine central peak. It measures 75 km (46 miles) across and 3.1 km (1.9 miles) deep. It's named for Simon Stevin (1548–1620), a Belgian mathematician and physicist.

21. Furnerius is another impressive walled plain, the southernmost of a great chain of four impressive craters south of Mare Crisium (A) – the others being Langrenus, Vendelinus, and Petavius – which dominate the southern regions of the young crescent. The plain measures 125 km (77.5 miles) across and 3.5 km (2.2 miles) deep at its northern end; the rest of the crater walls barely rise above the local terrain. Indeed, Furnerius is an old feature, worn and battered, with terraced walls and numerous impact scars, the largest being Furnerius B in the northern half, which has a central rise. The crater is named for French mathematician Georges Furner (unknown–1643).

22. Rheita is a nice impact feature 70 km (42 miles) across and 4.3 km (2.7 miles) deep with a sharply defined rim and terraced inner walls. It is named for Anton Maria Schyrle (c. 1597–1660) of Rhaetia, a Czechoslovakian astronomer and optician.

23. Vallis Rheita (Rheita Valley) is a fantastic lunar valley measuring 445 km (275 miles) in length. Its depth varies from 30 km (19 miles) at the northwest end and narrows to 10 km (6.2 miles) at the southeastern end.

24. Steinheil (left) / **Watt** (right) form a dynamic crater pair in the rugged southern lunar highlands when seen under the light of a low Sun. The craters date to the Nectarian period (3.9 to 3.85 billion years ago). Steinheil measures 67 km (41.5 miles) across and 3 km (1.9 miles) deep; Watt is nearly as wide as Steinheil but its depth is unknown. Steinheil is named for German astronomer and physicist Karl August von Steinheil (1801–1870); Watt is named for James Watt, a Scottish inventor (1736–1819).

25. Biela is prominent crater 76 km (47 miles) wide and 3.1 km (1.9 miles) deep in the southern highlands south-southwest of the dynamic duo Steinheil and Watt. Its walls are steep and terraced, and its floor is flat and has a complex central peak formation. The crater is named for Austrian astronomer Wilhelm von Biela (1782–1856).

Waxing crescent (2)

C. Lacus Somniorum (Lake of Dreams) is the largest "lake" (actually a plain) on the Moon, measuring 384 km (238 miles) across. It is, nevertheless, ironically, a somewhat inconspicuous plain of basaltic lava flows that requires strict attention to locate with certainty just north of the Taurus Mountains (see page 31).

D. Palus Somnii (Marsh of Sleep) is a patch of terrain, 143 km (89 miles) across, brighter than Mare Crisium to the east and Mare Tranquillitatis (see below) to the west, but darker, nonetheless, than the abutting cratered terrain to the east. If you catch the marsh under a low Sun, look for its characteristic "tan" coloration and the prominent impact crater, Lyell, on its western fringe, (seen just left of the label D).

E. Mare Tranquillitatis (Sea of Tranquillity) is a substantial "sea" measuring 873 km (541 miles) in diameter and is a vast plain of intermediate to young basaltic lavas from the Late Imbrium epoch, with numerous impact craters and other fine details. The lava's color appears bluish under a high Sun, a consequence, no doubt, of the high metallicity of the rock. Together with Mare Serenitatis (page 32) it forms the Man in the Moon's left eye.

F. Mare Nectaris (Sea of Nectar) is a stunningly dark lava plain 333 km (206 miles) that dates to the Nectarian period (3.9 to 3.8 billion years ago), when a massive impact event created the basin in which the sea resides. Just southwest of the sea's center is the 12-km- (7.4-mile-) wide Rosse crater. And under the right illumination, you can see part of a bright ray of ejecta debris slashing the floor from the southwest to the northeast.

a. Montes Taurus (Taurus Mountains) is a lumpish muddle of mountains 3 km (1.9 miles) high just south of Lacus Somniorum (C) and north of Mare Tranquillitatis (E). The 40-km- (25-mile-) wide crater Römer appears just to the right of the label a in the photograph. Note the shadow in this deep and steep-walled crater that plunges to a depth of 3.3 km (2 miles).

b. Montes Pyrenaeus (Pyrenees Mountains) is a small range of lunar mountains bordering the eastern edge of Mare Nectaris (F).

26. Hercules is a prominent crater marking the southeastern tip of Mare Frigoris (see page 34). It also lies just west of Atlas (3); the two are striking companions. The walls of Hercules are finely terraced and its floor quite detailed. The crater has been the site of numerous (and controversial) *Transient Lunar Phenomena* (TLPs) – unusual, and short-lived, activity, such as colored glows, flashes, obscurrations, or other abnormal changes on the lunar surface. Hercules is named for the Greek mythological Strongman.

27. Isidorus, with **Capella** to its upper right (northeast), is a fine crater pair just north of Mare Nectaris (F). Isidorus measures 42 km (26 miles) across and 1.6 km (1 mile deep) and has a prominent, bowl-shaped crater inside it near its western rim. Capella is a slightly larger crater to its east, measuring 29 km (30 miles) with an unknown depth; it's floor has substantial and irregular broad rise crossed (north-southeast) by the deep cleft Vallis Capella, which runs from the crater's north to its southeast rims. Isidorus is named for Roman astronomer St. Isidore of Seville (c. 570–636), while Capella is named for Spanish astronomer Martianus Cappela (c. 400–unknown).

28. Gutenberg is a conspicuous crater in a narrow spit of highland between Mare Fecunditatis (B) and Mare Nectaris (C). It measures 74 km (46 miles) across and 2.3 km (1.5 miles) deep. Its floor is irregular, and its northwest rim has been ruined by impacts. The crater is named for Johann Gutenberg (c. 1390–1468), the German inventor of the movable-type printing press.

29. Goclenius (see crater 12 on page 29).

30. Colombo with **Colombo-A** above it (see crater 13 on page 29).

31. Santbech (see crater 15 on page 29).

32. Fracastorius is a stunning bay-shaped remnant of an ancient lunar impact feature, which dominates the southern floor of Mare Nectaris. The crater spans an impressive 124 km (77 miles) across to an unknown depth. Its inward wall has been all but covered by once fluid lava flows, which welled up from Mare Nectaris and infiltrated the crater's floor; still standing tall, however, is Fracastorius's outer wall, which rises to a height of 2.4 km (1.5 miles). Under the right lighting conditions, those with large binoculars might see a slim rille running generally east to west across the crater's smooth floor. The crater is named for the Italian scholar, astronomer, and poet Girolamo Fracastoro (1483–1553).

33. Piccolomini is a beautiful circular depression 88 km (54.5 miles) across and 4.5 km (2.8 miles) deep. It was formed by an impact some 3.8 to 3.2 billion years ago during the Upper Imbrian epoch. Its floor is relatively smooth, and its walls slumped and terraced. The crater lies at the southwestern end of the impressive **Rupes Altai** (in shadow in the photo) – a 427 km- (265-mile) long escarpment, which is best seen during the waning gibbous phase. Piccolomini is named for Italian astronomer and Archbishop Alessandro Piccolomini (1508–1578).

34. Stiborius lies one crater diameter to Piccolomini's lower left (southwest) and is of a similar age. Stiborius measures 44 km (27.3 miles) across and 3.7 km (2.3 miles) deep. Its northeastern wall has been shattered by impact; the rest are terraced and surround an irregular floor with an offset central peak. The crater is named for Andreas Stoberl (1465–1515), a German astronomer and mathematician.

35. Metius and **Fabricius** (to its lower left, or southeast) are twin craters in the rough northern lunar highlands. They lead southward into Janssen (see below) and border the impressive gorge Vallis Rheita (see #23) to the east-northeast. Metius measures 88 km (54.6 miles) across and 3 km (1.9 miles) deep and has a relatively flat floor. It is named for Dutch astronomer Adriaan Adriaanszoon (1571–1635), who went by the name of Metius. Fabricius is 10 km (6.2 miles) smaller and 0.5 km (0.3 miles) shallower than Metius and has multiple central peaks. It dates to the Eratosthenian period, 3.2 to 1.1 billion years ago. The crater is named for German astronomer David Fabricius (Goldschmidt) (1564–1617).

36. Janssen is an extremely large and one of the Moon's most prominent impact features, especially when the Sun is rising over it. Janssen measures a whopping 190 km (118 miles) across with a depth of 2.9 km (1.8 miles). The crater lies in the southeastern region of the rugged northern lunar highlands, and its battered appearance belies that fact, showing several impact features within its confines, the largest of which is Fabricius (#35) to its northeast. The crater is named for French astronomer Pierre Jules Janssen (1824–1907).

37. Vlacq is a circular crater 89 km (55 miles) across and 3 km (1.9 miles) deep, which appears foreshortened due to its proximity to the Moon's eastern limb. It's a deep crater that has a level floor with a central hill. It's named for Dutch mathematician Adriaan Vlacq (c. 1600–1667). The near twin crater **Rosenberger** lies to its lower right (southeast). It's 7 km (4.3 miles) larger but 0.8 km (0.5 miles) shallower. The crater is highly eroded and has a nice impact crater Rossenberger D near its southern wall. It's named for German astronomer and mathematician Otto August Rosenberger (1800–1890).

First quarter

E1. Sinus Asperitatis (Bay of Roughness) is a 206-km- (128-mile-) long plain of flood basalt connecting Mare Tranquillitatis (E) to the north and and Mare Nectaris (F) to the southeast. Lunar highlands border it to the west and rough continental terrain contains it to the east.

G. Lacus Mortis (Lake of Death) is a pale lake 151 km (94 miles) in diameter, that lies within an almost indiscernible ring of some highly eroded impact rim. The prominent crater Bürg (#40) lies just east of the lake's center. The lake is also nestled between Lacus Somniorum (C) and the dark wash of easternmost section of Mare Frigoris (unlabeled here) to its north.

H. Mare Serenitatis (Sea of Serenity) is one of the Moon's most glorious seas. Nearly circular and bleakly smooth, it forms, together with Mare Tranquillitatis (E), the left eye of the Man in the Moon. Its lava floor dates to the Upper Imbrian period and breaches the rugged land to its northeast, where it drains into Lacus Somniorum (C). Under low illumination, the floor is just wild with fanciful rilles and wrinkle ridges (which rise only tens of meters above the floor), the most glorious of which is the **Dorsa Smirnov** (the stunning "Serpentine Ridge") – a fantastic 200-km- (124-mile-) long system of braided ridges (the largest on the Moon), which runs north to south near its its eastern border; when the Sun rises on it (between 5–8 days), there's no mistaking its slithering,

undulating form in binoculars. Just to the lower right of the label H in the photograph is the tiny but noticeable crater **Bessel** in the sea's southern half. The crater is only 16 km (9.9 miles) across and 1.7 km (1 mile) deep, but it seems to stand alone in silent isolation, like an island in a dark, watery grave. The crater stands out remarkably well because its rim is surrounded by a blanket of bright ejecta tossed out during the forming impact. Bessel also lies directly on another outstanding feature – one of the bright rays of ejecta that seems to originate (but not quite!) at the dynamic crater Tycho (# 109 on page 39). The ray slices the Sea in half running from the northeast to the southwest and appears most dramatic around the time of full Moon (see the photo on page 27).

c. Montes Caucasus (Caucasus Mountains) a prominent wedge of mountains 550 km (34 miles) long with some fantastic peaks lying between Mare Serenitatis (H) to the southeast and Mare Imbrium (not seen here in shadow) to the west. Some of the peaks are remarkable, the largest of which rises to an impressive 6 km (3.7 miles) above the surrounding terrain.

d. Montes Haemus (Haemus Mountains) are a shark's tooth range of mountains forming the southwestern border of Mare Serenitatis (H), the northeastern border of Mare Vaporum (K; in shadow here) and the fragmented northwestern border of Mare Tranquillitatus (E). The range is 560 km (341 miles) long and its tallest peak rises to an elevation of 2.4 km (1.5 miles). Just to the lower right of the label **d** in the photo is the crater **Menelaus** – a young crater, 27 km (16.7 miles) in diameter and an impressive 3 km (1.9 miles) deep. It is named for Menelaus of Alexandria (c. AD 98), a Greek geometrician and astronomer.

e. Rupes Altai (Altai Escarpment) is a 427-km- (265-mile-) long escarpment that terminates at the impressive crater Piccolomini (#33).

38. Aristoteles is a dramatic plain-filled crater just south of the gray wash of the easternmost section of Mare Frigoris (not labeled here) and makes an attractive pair with Eudoxus (see below). Aristoteles measures 87 km (53.9 miles) across and 3.3 km (2 miles) deep and is surrounded by bright ejecta and its uneven floor is dappled with low hills. It is named for the Greek astronomer and philosopher (383–322 BC).

39. Eudoxus is a relatively small – 37-km- (23-mile-) wide; 3.4-km- (2-mile-) deep – but prominent crater two crater diameters south of Aristoteles. To its south, look for the ruined ramparts of the ancient crater Alexander, which bleed into the Caucasus Mountains. It is named for the Greek astronomer (c. 408–355 BC).

40. Bürg is the most prominent feature in the demolished and flooded crater formation known as Lacus Mortis (G). The crater measures 40 km (25 miles) across and 1.8 m (1.1 miles) deep. Look for its substantial central peak. The crater is named for the Austrian astronomer Johann Tobias Bürg (1766–1834).

41. Posidonius is a shallow but remarkably fine walled plain 95 km (59 miles) across and 2.3 km (1.4 miles) deep. Its northern rim and exterior has been battered by impact craters and its floor is hilly with a central craterlet (Posidonius B) and low rille. It's named for Greek geographer Posidonius of Apamea (135?–51? BC).

42. Plinius stands boldly between Mare Serenitatis (H) to the northwest and Mare Tranquillitatis (E) to the southeast. It measures 43 km (27 miles) across and 4.3 km (2.7 miles) deep. Its walls are neatly terraced and its floor hilly with what appears to be double central craterlets. The crater is named for the Greek natural scientist Gaius Secundus (AD 23–79), better known as Pliny the Elder.

43. Julius Caesar is an extremely dark, lava-filled crater (one of the darkest on the Moon; perhaps owing to contrast effects), measuring 90 km (56 miles) across and 3.4 km (2.1 miles) deep. It's named for the Roman emperor (c. 102–44 BC).

44. Agrippa is a fine crater 46 km (28.5 miles) across and 3.1 km (1.9 miles) deep with its near twin **Godin** (about 10 km smaller but just as deep) below it. Agrippa is named for Greek astronomer Agrippa (unknown–AD 92). Godin is named for the French astronomer and mathematician Louis Godin (1704–1760).

45. Delambre is a conspicuous impact crater near the southwest border of Mare Tranquillitatis (E). It measures 52 km (32 miles) across and 3.5 km (2.1 miles) deep.

It has terraced walls, an irregular floor, and high and jumbled peaks. The crater's named for French astronomer Jean-Baptiste Joseph Delambre (1749–1822).

46. Abulfeda crater is nicely paired with **Almanon**, two crater diameters to the south-southeast. Both lie in the crater-rich lunar highlands and look timeworn and beaten. Abulfeda measures 62 km (38 miles) across and 3.1 km (1.9 miles) deep, while Almanon is 49 km (30 miles) across and 2.5 km (1.5 miles) deep. Note also **Tacitus** two crater diameters east-northeast of Almanon which marks the northwest end of Rupes Altai (e). Abulfeda is named for Syrian geographer Ismail Abu'l Fida (1273–1331). Almanon is named for Persian ruler and astronomer Abdalla Al Mamun. And Tacitus is named for the Roman historian Cornelius Tacitus (c. AD 55–120).

47. Theophilus is one of the finest craters on the Moon – an august impact feature 100 km (62 miles) across and 3.2 km (2 miles) deep, which stands prominently on the northwestern shore of Mare Nectaris (F). Overall its expansive floor is flat, although it is punctuated by a striking central mountain mass with three prominent peaks. It is named for Greek astronomer Theophilus of Alexandria, who died in AD 412. The solitary impact feature Mädler lies directly to the east of Theophilus on its rugged eastern rampart. Mädler measures 28 km (17 miles) across and 2.7 km (1.7) miles deep and has a low central peak and ridge. It is named for German astronomer Johann Heinrich Mädler (1794–1874).

48. Cyrillus is another interesting impact feature on the northwestern shores of Mare Nectaris (F). Its northeast rim has been erased by the impact crater Theophilus (#47), which overlaps it, making the two appear linked. Cyrillus measures 98 km (61 miles) across and 3.6 km (2.2 miles) deep. It has a worn central hill and a fine craterlet Cyrillus C. It is named for Egyptian theologian and chronologist Saint Cyril (unknown–AD 444).

49. Catharina marks the southern end of what is known as the Theophilus Chain, consisting of Theophilus (#47), Cyrillus (#48) and Catharina – a well worn and low-walled ring measuring 100 km (62 miles) across and 3.1 km (1.9 miles) deep; it is essentially the southern twin of Theophilus and has a dimpled flat floor. The three craters in the chain make a surprising sight in binoculars when the Sun is low to them. Catharina is named for Greek theologian and philosopher St. Catherine of Alexandria (c. 282–305).

50. Sacrobosco is a fantastic and highly worn crater west of the Rupes Altai escarpment (e) and a near match to either Catharina or Theophilus in size, measuring 98 km (61 miles) across and 2.9 km (1.8 miles) deep. It's easily identified by the three craterlets on its otherwise flat floor. The crater is named for John of Hollywood, Johannes Sacrobuschus (c. 1200–c. 1256), a British astronomer. The two deep and shadowy craters to its upper left are **Geber** and **Abenezra**; these are part of another nice three-crater chain of similarly sized craters, with Almanon (see #46) to the northeast. Geber measures 45 km (28 miles) across and 3.5 km (2.2 miles) deep;

it is named for Spanish-Arab astronomer Gabir Ben Aflah Geber (unknown–c. 1145). Abenezra is 3 km smaller and about as deep; it is named for Abraham Bar Rabbi BenEzra (c. 1092–1167), a Spanish-Jewish mathematician and astronomer.

51. Zagut is another worn crater in the rugged, southern lunar highlands. It measures 84 km (52 miles) across and 3.2 km (1.9 miles) deep. Its eastern wall is ruined by the elliptical crater Zagut E, which links to **Lindenau** crater further to the east. Zagut is named for Spanish-Jewish astronomer Abraham Ben Samuel (c. 1450–1522). Lindenau measures 53 km (33 miles) across and 2.9 km (1.8 miles) deep; it is named for German astronomer Bernhard von Lindenau (1780–1854).

52. Gemma Frisius is a fine sight with a prominent impact scar **Goodacre** overlapping its northeastern rim. Gemma Frisius measures 88 km (55 miles) across and 4.7 km (2.9 miles) deep. Its north and east rim and ramparts have also been scarred by smaller impacts, giving it the distinct appearance of a paw-print. The crater's floor is rugged and has a small central rise. It is named for Dutch doctor, mathematician, and cartographer Jemma Reinier (Gemma Frisius) (1508–1555). Goodacre measures 46 km (28.5 miles) across and 3.2 km (2 miles) deep. It too is heavily eroded and damaged by impacts. It's named for British selenographer Walter Goodacre (1856–1938).

53. Maurolycus is a great crater that abuts **Barocius** to its southeast. Maurolycus measures an impressive 114 km (71 miles) across and 4.7 km (2.9 miles) deep. Its walls are fractured and broken by impacts and its floor is rough with ruined rings and has a central rise. It is named for Italian mathematician Francesco Maurolico (1494–1575). Barocius measures 82 km (51 miles) across and 3.5 km (2.2 miles) deep. It too is well worn and suffered a major impact on its northeast rim (Barocius B); it is named for Francesco Barocius, an Italian mathematician (1494–1575). To the southwest of Barocius lies the equally impressive (and equally worn and battered) **Clairaut** crater, which measures 75 km (46.5 miles)

across and 2.7 km (1.7 miles) deep. It has a pretty Y-shaped cluster of craters to its lower right, the last of which is Baco. Clairaut is named for French mathematician Alexis Claude Clairaut (1713–1765).

54. Cuvier is a highly eroded impact crater in the deep southern highlands. It measures 75 km (46.5 miles) across and 3.8 km (2.4 miles) deep. It is named for French natural scientist, paleontologist Georges Cuvier (1769–1832).

55. Jacobi is another moderately sized and well worn and battered impact crater in the deep southern highlands. It measures 68 km (42 miles) across and 3.3 km (2 miles) deep. It is named for German mathematician Karl Gustav Jacob (1804–1851).

56. Pitiscus is a pretty crater with narrow walls and a pitted floor that was flooded by lava. It measures 82 km (51 miles) across and 3 (1.9 miles) deep and has a low central peak. It is named for Bartholemaeus Pitiscus (1561–1613), a German mathematician.

57. Hommel is a deep southern crater just west of Rosenberger (#37) and south of Pitiscus – a striking crater triad. The crater is close enough to the lunar limb that its round form appears foreshortened. It measures 120 km (74 miles) across and 2.8 km (1.7 miles) deep. Most prominent are the large impact scars that ruin its walls, including Hommel H to the northwest, Hommel B in the east, and Hommel P in the southern wall. It is named for German astronomer and mathematician Johann Hommel (1518–1562).

58. Manzinus is a far southern circular crater that appears foreshortened due to its proximity to the southern limb. It measures 98 km (61 miles) across and 3.8 km (2.4 miles) deep. It's named for Italian astronomer Carlo Antonio Manzini (1599–1677). Manzinus is nicely paired with **Mutus** one crater diameter to its northeast. Mutus is deep and its walls are highly eroded; its floor, however, appears as a smooth plain. The crater measures 78 km (48 miles) across and 3.7 km (2.3 miles) deep. It's named for Spanish astronomer and explorer Vincente Mut, or Muth (unknown–1673).

Waxing gibbous

I. (East) Mare Frigoris (Sea of Cold) is one of the Moon's long (1,596 km; 989 miles), irregular seas. Only the eastern part of it is seen here near the Moon's northern limb, just north of Mare Serenitatis (H). The lavas in this eastern section date to the Late Imbrian epoch (~3.8 to 3.2 billion years ago). They differ from those in Western Mare Frigoris (not seen here), which date to the Eratosthenian period (~3.2 to 1.1 billion years ago).

J. Palus Putredinis (Marsh of Decay) is a small patch of lava-flooded plain – 161 km (100 miles) across – between Archimedes crater (#60) and Montes Apenninus (g).

K. Mare Vaporum (Sea of Vapors) is a 245-km-wide (152-mile-wide) sea southwest of Mare Serenitatis (H)

and southeast of the Montes Apenninus (g). Its lava date to the Eratosthenian period (~3.2 to 1.1 billion years ago). It forms the Lady in the Moon's eye.

L. Sinus Medii (Central Bay) is a small patch of mare [335 km (208 miles)] southwest of Mare Vaporum. It forms the mouth of the Lady in the Moon.

O. Mare Imbrium (see page 37).

f. Montes Alpes are named for the European Alps and form a lunar mountain range that stretches across 180 km (112 miles) of the Moon's surface, forming the northeastern border of Mare Imbrium (see page 37). The highest peaks in the range rise 1.8 to 2.4 km (1.2 to 1.5 miles) above the surrounding mare.

g. Montes Apenninus (Apennine Mountains) form a lovely curved mountain range that begins at the prominent crater Eratosthenes (not seen here; see #96) and

extends 600 km (372 miles) to the northeast. Its peaks rise to a maximum altitude of 5 km (3 miles); they form the southeastern border of Mare Imbrium).

h. Rupes Recta (Straight Wall), seen here mostly in shadow, is one of the Moon's most treasured geological features: a tectonic fault 110 km (68 miles) long and only 300 m (984 feet) at its highest extent. The fault is not truly a wall (more of a gentle slope) 2.5 km (1.5 miles) wide. The face of the "wall" appears as a dark line during the waxing gibbous phases of the Moon and as a white line during its waning gibbous phases.

59. Aristillus and **Autolycus** (two crater diameters to its south-southwest) are an attractive crater pair in eastern Mare Imbrium (O; see page 37), and make an even finer crater triad with Archimedes (#60) further to the southwest. Aristillus measures 55 km (34 miles) across and 3.6 km (2.2 miles) deep. Its inner walls are neatly terraced and it has a beautiful central cluster of peaks. It is named for Greek astronomer Aristillus (c. 280 BC). Autolycus measures 39 km (24 miles) across and 3.4 (2.1 miles) deep with no central peak. It is named for Greek astronomer Autolycus (unknown– c. 310 BC).

60. Archimedes is the largest crater on Mare Imbrium (O; see page 37) and looks like a walled plain with its shallow rise and flat lava-filled floor. The crater measures 83 km (51.5 miles) across and 2.1 km (1.3 miles) deep and is surrounded by a short rampart formed by ejecta. It is named for Greek physicist and mathematician Archimedes (c. 287–212 BC).

61. Manilius is a small but bright-walled crater in northeastern Mare Vaporum (K). It measures 39 km (24 miles) across and 3.1 km (1.9 miles) deep and has a small central peak. It is named for Roman writer Marcus Manilius (unknown–c. 50 BC).

62. Pallas is an extremely eroded crater with a low wall and lava-flooded floor. It measures 50 km (31 miles) across and only 1.3 km (0.8 miles) deep. It is named for German geologist and natural historian Peter Simon Pallas (1741–1811).

63. Triesnecker is a small but noticeable crater in the lava plain between Mare Vaporum (K) and Sinus Medii (L). It measures 28 km (17.4 miles) across and 2.8 km (1.7 miles) deep. It is named for Austrian astronomer Francis A. Paula Triesnecker (1745–1817).

64. Flammarion is a well-worn and ruined crater in southern Sinus Medii. It measures 75 km (46.5 miles) across and only 1.5 km (0.9 miles) deep. Its floor has been flooded with lava. It is named for the French astronomer and astronomy popularizer Camille Flammarion (1842–1925). The beautiful crater **Herschel** lies to its southeast. It measures 41 km (25 miles) across and 3.8 (2.4 miles) deep. It has a nice circular form, terraced walls, and a large central rise. It is named for the German-born British astronomer Sir William Herschel (1738–1822). Large-binocular viewers should also look for the fully flooded crater **Spörer** to its north.

65. Ptolemaeus is an enormous walled plain neatly situated near the Moon's center in the central highlands. It measures a whopping 153 km (95 miles) across and only 2.4 km (1.5 miles) deep. Its floor is fantastically flat, smooth, and eerily dark (no doubt a contrast effect). It is the northernmost member of a three-crater chain with Alphonsus (#68) and Arzachel (#69). It is named for Greek mathematician, astronomer, and geographer Claudius Ptolemy (AD 87–150).

66. Hipparchus is a magnificent structure 150 km (93 miles) across and 3.3 km (2 miles) deep. It's extremely eroded and battered and difficult to see under high-Sun illuminations. But it's a fantastic sight seen under low sunlight. It's named for Greek astronomer Hipparchus (unknown–140 BC). The prominent crater **Horrocks** is near the northeastern rim of Hipparchus.

67. Albategnius is another large walled plain in the Moon's central highlands, just south of Hipparchus. It measures 129 km (80 miles) across and 4.4 km (2.7 miles) deep and is highly detailed, having impact scars, landslides, and valleys. It is named for the Iraqi astronomer Al-Batani (c. 858–929). The crater **Halley** is the prominent impact feature just northeast of its northern lip and the smaller **Hind** lies to the right of Halley.

68. Alphonsus is prominent crater in the central lunar highlands and the middle of the great Ptolemaeus chain. It measures 119 km (74 miles) across and 2.7 km (1.7 miles) deep. The floor is most interesting, being laced with long rilles and dark-collared cinder cone-shaped craters. This crater caused quite a stir in the mid-to-late 1950s when transient lunar phenomena were recorded in it. The most famous account came from Soviet astronomer Nikolai Kozyrev, who observed a mist-like obscuration on the crater's floor, which he took to be a sign of volcanic-related activity. The crater is named for the Spanish astronomer Alfonso the 10th (The Wise One), King of Castile (1221–1284). **Alpetragius** is the finely sculpted impact crater abutting it to the southwest.

69. Arzachel is the southern member of the great Ptolemaeus chain. It measures 96 km (59.5 miles) across, and 3.6 km (2.2 miles) deep and has youthful walls, a flat floor, prominent rille, and a central mountain. It is named for Spanish-Arabic astronomer Al Zarkala (c. 1028–1087).

70. Thebit is the bright crater just east of the Straight Wall (h). Thebit measures 57 km (35 miles) across and 3.3 km (2 miles) deep and has an impact scar (Thebit A) on its west-northwest rim. It is named for Iraqi astronomer Ben Korra (Thabit ibn Qurra) (836–901).

71. Purbach is a substantial walled plain 118 km (73 miles) across and 3 km (1.9 miles) deep in the south central highlands. Its walls are heavily worn and its floor is generally smooth with no obvious central peak. It is the northernmost member in the Walter (#77) chain of craters with **Regiomontanus** to its south. Regiomontanus is equally impressive, measuring 126 × 110 km (78 × 68 miles) across and a shallow 1.7 km (1 mile) deep. Like Purbach, its walls are heavily eroded and its floor lava smooth with some peaks offset from the crater's center. Near the crater's northwest wall is **Regiomontanus A**, whose central peak has been the site of lunar transient phenomena in the past. Purbach is named for Austrian mathematician Georg von Purbach (1423–1461). Regiomontanus is named for German astronomer and mathematician Regiomontanus or Johann Muller (1436–1476).

72. Blanchinus is an irregular crater in the south central highlands. It measures 68 km (42 miles) across and 4.2 km (2.6 miles) deep. Its walls are worn and its floor flat. The crater is named for Italian astronomer Giovanni Bianchini (unknown–1458). The crater **La Caille** is linked to its northwest rim. It measures 68 km (42 miles) across

and 2.8 km (1.7 miles) deep and has a flat lava-filled floor. It is named for French astronomer Nicholas Louis De La Caille (1713–1762).

73. Apianus is another worn and smooth-floored crater in the south central highlands. It measures 63 km (39 miles) across and 2 km (1.2 miles) deep. The crater **Playfair** is two crater diameters to the northwest. Apianus is named for German mathematician and astronomer Peter Bienewitz (Petrus Apianus) (1495–1552).

74. Azophi is a moderately sized crater linked to its twin **Abenezra** to its northwest. Azophi measures 47 km (29 miles) across and 3.7 km (2.3 miles) deep. Abenezra is just as deep and only 5 km (3 miles) smaller. Azophi is named for Persian astronomer Abderrahman Al-Sufi (903–986). Abenezra is named for Spanish-Jewish mathematician and astronomer Abraham Bar Rabbi BenEzra (c. 1092–1167).

75. Geber is a moderately sized crater similar in size to Azophi (#74) and Abenezra. It is named for Spanish Arab astronomer Gabir Ben Aflah (unknown–c. 1145).

76. Aliacensis is a grand lunar crater, wonderfully circular, measuring 79 km (49 miles) across and 3.7 km (2.3 miles) deep. Its floor is flat and has a gentle central rise. It's named for French geographer Pierre D'Ailly (1350–1420). The twin crater **Werner** lies just to the northwest. It is named for German mathematician Johann Werner (1468–1528).

77. Walter is a noble walled plain 130 × 140 km (81 × 87 miles) across and 4.1 km (2.5 miles) deep in the rugged southern highlands. While its borders are worn its floor is relatively smoothed with a central peak slightly offset from center. It's named for German astronomer Bernard Walter (Walther) (1430–1504).

78. Deslandres is a fantastic impact scar 256 km (159 miles) across of unknown depth. It is the largest impact crater on the Moon that's easily accessible in binoculars away from the limb. The floor is relatively smooth, overall, and is home to **Cassini's Bright Spot** – one of the brightest features on the full Moon. The walls have been excessively worn and battered, and the crater sports two substantial impact features: **Hell** crater lies entirely within its western rim, and **Lexell** crater intrudes on its southeastern wall. Deslandres is named for French astrophysicist Henri Alexandre Deslandres (1853–1948).

79. Orontius crater is an irregular impact feature in the southern highlands. It measures 122 km (76 miles) across and 3.1 km (1.9 miles) deep. It's another eroded and battered crater with a very interesting chain of three overlapping craters (**Huggins–Nasireddin–Miller**) extending to the east and north. Orontius is named for French mathematician and cartographer Finnaeus Orontius Oronce Fine (1494–1555).

80. Saussure crater is one crater diameter south (and slightly east) of Orontius (#79). It measures 54 km (33 miles) across and 1.8 km (1.1 miles) deep. It is named for Swiss geologist Horace Benedict De Saussre (1740–1799).

81. Stöfler is another worn and battered southern highland crater with an impressive diameter of 126 km

(78 miles) and a depth of 2.8 km (1.7 miles). Its floor is relatively dark, making it easy to spot in the highly reflective surroundings. **Fernelius** borders it to the north and **Faraday** lies to the southeast. The crater is named for German astronomer and mathematician Johann Stöfler (1452–1531).

82. Licetus is joined to the southwest by the large irregular crater formation **Heraclitus**. Licetus measures 75 km (46.5 miles) across and 3.8 km (2.4 miles) deep. Its floor is flat and its southern wall has crumbled into its southern partner. It is named for Italian physicist, philosopher, and doctor Fortunio Liceti (1577–1657).

83. Maginus is another southern lunar highland wonder, measuring an impressive 194 km (120 miles) across and 4.3 km (2.7 miles) deep. Its walls are uneven and its

floor flat with some low central peaks. The crater **Proctor** lies to its northeast, and its southeastern rampart has been smashed by the now eroded crater **Maginus C**. Maginus is named for Italian astronomer and mathematician Giovanni Antonio Magini (1555–1617).

84. Moretus is a highly foreshortened southern highland crater near the Moon's south pole. It measures an impressive 114 km (71 miles) across and 5 km (3 miles) deep! Its central peak is magnificent, rising to 2.1 km (1.3 miles) above the crater's smooth floor and may just be the largest central peak on the Moon. **Curtius** crater borders it to the northeast, while crater **Short** lies to its south. Moretus is named for Belgian mathematician Theodore Moret (1602–1667).

Last quarter

I. (West) Mare Frigoris (Sea of Cold) is part of one of the Moon's long (1,596 km; 989 miles), irregular seas (see page 34).

M. Sinus Roris (Bay of Dew) is separated from Mare Frigoris to the east by a narrow "marsh" of rougher terrain. It is essentially the northern extension of the vast Oceanus Procellarum (see page 38). It measures 202 km (125 miles) across.

N. Sinus Iridum (Bay of Rainbows) is a striking crescent bay of basaltic lava that forms the northwestern brow of Mare Imbrium (O). It measures 236 km (146 miles) across.

O. Mare Imbrium (Sea of Rains) is a vast lava-filled basin – the greatest circular sea on the Moon. It forms the Man in the Moon's right eye and measures 1,123 km

(696 miles) across and bleeds into the beautiful crescent-shaped bay of Sinus Iridum. The basin itself is marked by three concentric mountain ranges: Montes Alpes (f) to the northeast, Montes Carpatus (n) to the south, and Montes Apenninus (g) to the southeast.

P. Sinus Aestuum (Seething Bay) is a dark and largely featureless plain east of Copernicus crater (97) and south of the Apennine Mountains (g). It is separated from Mare Vaporum (K) to the east by a neck of jumbled terrain. It measures 290 km (180 miles) across.

Q. Mare Cognitum (Sea That Has Become Known) is a minor elliptical patch of flood basalt 376 km (233 miles) across. It connects to Oceanus Procellarum (S) to the west, Mare Nubium (R) to the southeast, and Mare Humorum (T) to the southwest.

R. Mare Nubium (Sea of Clouds) is a large lunar sea, 715 km (443 miles) across, just west of the low central

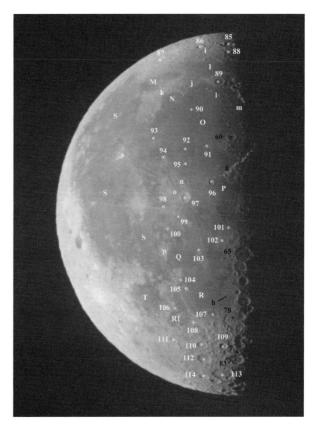

lunar highlands. The lavas here date to the Late Imbrian epoch between 3.8 and 3.2 billion years ago.

R1. Palus Epidemiarum (Marsh of Diseases) is an east–west trending dash of lunar basalt just southwest of Mare Nubium (R). It measures 286 km (177 miles) across, is irregular in form, and laced with rilles.

S. Oceanus Procellarum (Ocean of Storms) is an enormously long – 2,568 km (1,592 miles) – and wide-shaped crescent sea on the western side of the Moon and forms the Man in the Moon's mouth. It is bordered to the southeast by Mare Cognitum (Q) and Mare Humorum (T), and to the northeast by Mare Imbrium (O) and to the north by Sinus Roris (M). Its entire surface is comprised of flood basalt which has been scarred in places by prominent impacts.

T. Mare Humorum (Sea of Moisture) is a lava-flooded impact basin 389 km (241 miles) near the southwestern limb of the Moon. The depth of the circular basin is believed to be 3 km (1.9 miles) deep and its surface is fractured and rimmed by large impacts.

i. Philolaus Escarpment is a deep cliff-like feature in the northern highlands just east of Philolus crater (#86).

j. Plato Uplands is a broken crescent of irregular terrain between the prominent crater Plato (#89) and the Jura Mountains (k) surrounding Sinus Iridum (N).

k. Montes Jura (Jura Mountains) is one of the most striking semicircular formations on the Moon. The mountains curl around Sinus Iridum (N) to the northwest to an extent of 422 km (261 miles). The mountains are the remains of a crater 260 km (161 miles) in diameter that was flooded with basaltic lava.

l. Montes Teneriffe (Teneriffe Mountains) are a 110-km- (68-mile-) long and scattered range of mountains in the northern part of Mare Imbrium (O), southwest of crater Plato (#89).

m. Mons Piton (Mount Piton) is a bright, isolated mountain peak in eastern Mare Imbrium (O) rising 2.3 km (1.4 miles) above the surrounding lava plain.

n. Montes Carpatus (Carpathian Mountains) are a discontinuous mountain range forming the southern border of Mare Imbrium (O). The semi-rugged range extends 361 km (224 miles) and has peaks rising to 2,134 m (7,000 feet).

o. Carpathian Plateau is an unofficial region referring to the bright ejecta west of the crater Copernicus. The name first appeared on the 1969 Rand McNally Map of the Moon.

p. Montes Riphaeus (Riphaeus Mountains) is a jumbled patch of rugged peaks defining the west-northwest border of Mare Cognitum (Q) at the southeastern end of Oceanus Procellarum (S). The mountain range, which is oriented north–northeast to south–southwest, measures some 190 km (118 miles) in length and 50 km (31 miles) in width.

85. Goldschmidt is a large and lovely walled plain in the northern lunar highlands. It measures 113 km (70 miles) across and 2 km (1.2 miles) deep. Its walls appear rough and ragged and its western rim has been ruined by the Anaxagoras impact. The floor is relatively smooth but riddled with fine details at low Sun elevations. It is named for German astronomer Hermann Goldschmidt (1802–1866).

86. Philolaus is an interesting crater three crater diameters west of Goldschmidt (#85) and is separated from it by the equally fascinating Philolaus Escarpment (i). It measures 70 km (43 miles) across and is 3.4 km (2.1 miles) deep. Its floor is irregularly detailed with scattered peaks. It is named for the Greek mathematician, astronomer, and philosopher Philolaus of Croton (c. 480–c. 385 BC).

87. Pythagoras is a grand crater on the western edge of the northern lunar highlands. Its walls appear to form a large ellipse but this is due to foreshortening. The crater is an impressive 130 km (80 miles) across and plunges to depths of 5 km (3 miles). Most prominent are its twin central peaks, which rise 1.5 km (0.9 miles) above the crater's gently undulating floor. It is named for the Greek philosopher and mathematician Pythagoras of Samos (unknown–c. 532 BC).

88. Epigenes is an interesting crater due south of Goldschmidt (#85). It measures only 55 km (34 miles) across and 2 km (1.2 miles) deep. Its north and northwest walls are well formed but the remaining rim is ruined. Its floor is equally complex. It is named for Greek astronomer Epigenes (unknown–c. 200 BC).

89. Plato is one of the most obvious great walled plains – Hevelius's "Greater Black Lake" – in the lunar Alps, just north of Mare Imbrium (O). Its wonderfully circular form measures 109 km (64 miles) across and 1 km (0.6 miles) deep. The floor is remarkably smooth and flat, filled with flood basalt, and has no central crater. The floor has peculiar irregular shades that seem to change with shifting Sun angles. Not surprisingly, it has been the site of numerous transient lunar phenomena. Plato is only slightly younger than Mare Imbrium dating to about 3.84 billion years ago. It is named for Greek Philosopher Plato (c. 428–c. 347 BC).

90. Le Verrier is a tiny lunar crater only 20 km (12.4 miles) across and 2.1 km (1.3 miles) deep on the northwestern floor of Mare Imbrium (O). It has a slightly larger twin, **Helicon** crater, just to its west. Le Verrier is named for the French mathematician Urbain Le Verrier (1811–1877) who helped discover Neptune.

91. Timocharis is another small but noticeable impact crater on the vast floor of Mare Imbrium (O). It measures 34 km (21 miles) across and 3.1 km (1.9 miles) deep. Its walls are finely terraced and its ramparts bright and wide, fanning out 20 km (12.4 miles) across the dark sea in all directions. Indeed, it is the core of a dim ray system. The crater also has a fine central peak. The crater is named for Greek astronomer Timocharis (unknown–c. 280 BC).

92. Lambert is another small but apparent impact crater on the floor of Mare Imbrium (O). It measures 30 km (19 miles) across and 2.7 km (1.7 miles) deep. It has a bright rampart with ray-like extensions, a terraced wall, and a central craterlet. It is named for German astronomer Johann Heinrich Lambert (1728–1777).

93. Delisle is a small crater in western Mare Imbrium (O) measuring 25 km (15.5 miles) across and 2.6 km (1.6 miles) deep, surrounded by a small, rippled rampart. **Montes Delisle** (actually a high ridge) lies to its southwest, while the smaller crater **Diophantus** lies to its south-southeast (not visible in this photo, but becomes apparent under low Sun angles). Delisle is named for French astronomer Joseph Nicolas Delisle (1688–1768).

94. Euler is a small crater on the southwestern floor of Mare Imbrium (O) bordered to the southwest by a series of ridges (again not visible in this photo, but become apparent under low Sun angles). It measures 28 km (17 miles) across and 2.2 km (1.4 miles) deep, with a small system of rays. It is named for Leonhard Euler (1707–1783), a Swiss mathematician.

95. Pytheas is another small but noticeable impact crater on the southern floor of Mare Imbrium (O). It measures 20 km (1.2 miles) across and 2.5 km (1.5 miles) deep. It is named for the fourth-century BC Greek navigator and geographer Pytheas of Marseilles.

96. Eratosthenes is a small but deep impact crater punctuating the southern end of the lunar Apennine Mountains (g). It measures 58 km (36 miles) across and 3.6 km (2.2 miles) deep. At low Sun angles, its high rim casts a marvelous shadow across the neighboring flatlands. It may have been formed about 3.2 billion years ago, and in a 1924 *Popular Astronomy* article, Harvard astronomer William Pickering believed in the perhaps fanciful and frail notion that the slowly moving shadows playing on the crater's jumbled floor were being cast by migrating animals. Rays from the magnificent crater Copernicus (#97) splash across Eratosthenes. Eratosthenes is named for the Greek astronomer and geographer Eratosthenes (c. 276–196 BC).

97. Copernicus – "Monarch of the Moon" – is arguably the grandest crater formation available to us on the Moon's near side. Centrally located, this 800-million-year-old impact feature stands boldly against the Moon's vast dark central seas, surrounded by a brilliant ejecta blanket with needle-like rays that extend 800 km (496 miles) across the surrounding plains. The walls are wonderfully terraced and its floor graced with a complex central peak that rises 1.2 km (0.7 miles) above the crater's floor. The crater is named for Polish astronomer Nicholas Copernicus (1473–1543).

98. Hortensius is a tiny (but noticeable) crater 15 km (9 miles) across and 2.9 km (1.8 miles) deep west-southwest of Copernicus (#97). It is named for Dutch astronomer Martin van den Hove (1605–1639).

99. Reinhold is a moderately large impact scar south-southwest of Copernicus (#97). It measures 48 km (30 miles) across and 3.3 km (2 miles) deep. It is named for German astronomer and mathematician Erasmus Reinhold (1511–1553).

100. Lansberg is a small crater south-southwest of Reinhold (#99) and Copernicus (#97). It measures 39 km (24 miles) across and 3.1 km (1.9 miles) deep. It is named for Belgian astronomer Philippe van Lansberg (1561–1632).

101. Mösting is a small but noticeable impact crater near the central highlands. It measures 26 km (16 miles) across and 2.8 km (1.7 miles) deep. The large teardrop-shaped walled plain **Flammarion** lies to its southeast. Mösting is named for Danish benefactor Johan Sigismund Von Mösting (1759–1843).

102. Lalande is a small impact scar southwest of Mösting and Flammarion. It measures 24 km (15 miles) across and 2.6 km (1.6 miles) deep. The crater may be 2.8 billion years old. It is named for French astronomer Joseph Jerome Le Francois De LaLande (1732–1807).

103. Fra Mauro is an old and ruined walled plain on the northeastern border of Mare Cognitum (Q). The crater measures in impressive 95 km (60 miles) across, but its walls have all but been eroded into the surrounding plains. It is named for Italian geographer Fra Mauro (unknown–1459).

104. Lubiniezky is a small lava-filled crater on the northwest rim of Mare Nubium (R). It measures 44 km (27 miles) across and 0.8 km (0.5 miles) deep. It is named for Polish astronomer Stanislaus Lubiniezky (1623–1675).

105. Bullialdus is a moderately sized crater southeast of Lubiniezky in the Mare Nubium (R). The crater measures 61 km (38 miles) across and 3.5 km (2.2 miles) deep. It is named for French astronomer Ismael Boulliaud (1605–1694).

106. Campanus is another small lunar impact crater in the small lunar sea Palus Epidemiarum (R1). It measures 48 km (30 miles) across and 2.1 km (1.3 miles) deep. The equally sized crater **Mercator** is immediately to its southeast. Campanus is named for Italian astronomer Giovanni Campano (c. 1200–1296). Mercator is named for Belgian cartographer Gerard De Kremer (Gerhardus Mercator) (1512–1594).

107. Pitatus is a large walled plain with extremely worn walls and a lava-flooded floor (a ghost crater). It measures 97 km (60 miles) across and only 0.9 km (0.6 miles) deep. **Wurzelbauer** lies to its southwest; **Gauricus** is just to the south-southeast. Pitatus is named for Italian astronomer and mathematician Pietro Pitati (unknown–1550).

108. Chicus is a small crater in the southern highlands west of Wurzelbauer and separated from it by one of the prominent rays emanating from Tycho crater (#109). It is named for the thirteenth-century Italian astronomer Franceso Degli Stabili.

109. Tycho is a stunningly bright and sharply chiseled crater prominently placed in the lunar southern highlands. It is most famous for being the center of a bright and extensive system of rays (the greatest on the Moon), which radiate from Tycho clear across parts of the Moon's face. Through binoculars, Tycho marks the position of the pearl in the Lady in the Moon's necklace (two of Tycho's rays), which is especially prominent during the full Moon. Interestingly, to the unaided eyes, the "pearl" is not Tycho, but Cassini's Bright Spot in the neighboring crater Deslandre (#78). Tycho measures 85 km (53 miles) across and an impressive 4.8 km (3 miles) deep. The crater has an estimated age of 108 million years young and its steep, terraced walls surround a dimpled floor with a strong central peak. The crater is named for Danish astronomer Tycho Brahe (1546–1601).

110. Wilhelm is a large prominent walled plain, about three crater diameters west of Tycho in the rugged southern lunar highlands. The crater measures 107 km (66 miles) across and 3 km (1.9 miles) deep. Ejecta from Tycho cut across the crater's smooth floor, which has some lofty peaks. It is named for German astronomer Wilhelm IV (Landgrave of Hesse–Kassel) (1532–1592).

111. Hainzel is a prominent member of a large group of overlapping craters in the western part of the southern highlands. It measures 70 km (43 miles) across and 3 km (1.9 miles) deep. It is named for German astronomer Paul Hainzel (1527–1581).

112. Longomontanus is a large and obvious walled plain southwest of Tycho in the southern lunar highlands. It measures 145 km (90 miles) across and 4.5 km (2.8 miles) deep. Its walls are worn and its rim is level with the surrounding highlands. The floor is flat but detailed under low Sun angles, with a peak complex.

It is named for Danish astronomer Christian Sorensen Longomontanus (1562–1647).

113. Clavius is one of the largest crater formations on the Moon and lies in the deep southern highlands, where its flat and pitted floor stands out prominently from the rugged surroundings. The crater measures 225 km (139.5 miles) across and 3.5 km (2.2 miles) deep and appears slightly elongated due to libration effects. Its walls are battered and crumbling, and its floor is greatly pitted, with a beautiful curving chain of significant impact scars with ever-diminishing diameters. The crater is in fact one of the older formations, dating to about four billion years ago. The prominent crater **Blancanus** (which itself is an impressive 105 km (65 miles) across and 3.7 km (2.3 miles) deep) touches it to the southwest. Clavius is named for German mathematician Christopher Klau (1537–1612). Blancanus is named for Italian mathematician and astronomer Giuseppe Biancani (1566–1624).

114. Scheiner is another large-walled plain (though it seems small when compared to Great Clavius) to the west-southwest of Clavius. It measures 110 km (68 miles) across and 4.5 km (2.8 miles) deep, making it a near twin of Blancanus to its southeast. It is also heavily eroded and pitted, especially to the north. It is named for German astronomer Christopher Scheiner (1573–1650).

Waning crescent

q. Montes Harbinger (Harbinger Mountains) are a sparse but remarkable gathering of lofty peaks northeast of Aristarchus (#120), which connect to the broken walled plain (a slender crescent) **Prinz**.

115. J. Herschel is a large and highly foreshortened walled plain in the lunar uplands bordering Mare Frigoris. It measures a substantial 165 km (102 miles) across but is of unknown depth; its walls appear to have been all but eroded away. It is named for British astronomer John Herschel (1792–1871).

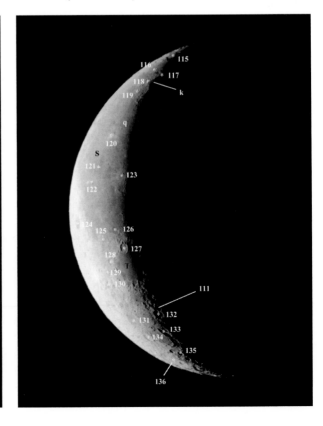

116. Harpalus is a small and young impact crater on the eastern edge of Sinus Roris. It measures 39 km (24 miles) across and 2.9 km (1.8 miles) deep. It is named for Greek astronomer Harpalus (unknown–460 BC).

117. Bianchini is a small crater at the northern end of the Jura Mountains (k). It measures 38 km (24 miles) across and 3.1 km (1.9 miles) deep. It's named for Italian astronomer Francesco Bianchini (1662–1729).

118. Sharp is a bold crater in the Jura Mountains (k) hugging Sinus Iridum. It measures 40 km (25 miles) across and 3.2 km (2 miles) deep and has a low hill at its center. It is named for British astronomer and mathematician Abraham Sharp (1651–1742).

119. Mairan is a small but noticeable impact crater on a highland peninsula between Oceanus Procellarum (S) and the Mare Imbrium (O). It measures 40 km (25 miles) across and 3.4 (2.1 miles) deep. It is named for French geophysicist Jean Jacques D'Ortous De Mairan (1678–1771).

120. Aristarchus is truly one of the great wonders on the near side of the Moon. It is the brightest formation on the near side of the Moon; it is so intense that it dominates the view when the young crescent Moon is bathed in earthshine. Seeing it in this way led William Herschel to believe it was a volcano erupting on the Moon's surface. Not surprisingly, it continues to be a site for sightings of transient lunar phenomena. Interestingly, though, it is a site of many lunar volcanic features including **Vallis Schröteri (Schröter's Valley)** – the largest sinuous rille on the Moon, which starts on the Aristarchus plateau (just north of the walled plain **Herodotus**, which lies immediately to the east-southeast of Aristarchus.) Aristarchus measures 40 km (25 miles) across and 3.7 km (2.3 miles) deep. It is named for Greek astronomer Aristarchus (310–230? BC).

121. Marius is a noticeable crater on the floor of Oceanus Procellarum. It measures 41 km (25 miles) across and 1.7 km (1 mile) deep. It is named for German astronomer Simon Mayer (Marius) (1573–1624).

122. Reiner is a small but noticeable crater centrally located on the vast floor of Oceanus Procellarum. It measures 30 km (19 miles) across and 2.6 km (1.6 miles) deep. It is named for Italian astronomer Vincentio Reinieri (1606–1647).

123. Kepler is another young and majestic impact formation on the near side of the Moon, between Copernicus crater (#97) to the east and Oceanus Procellarum to the west. It has a bright rampart with long rays that tickle the surrounding seas to extents of 300 km (186 miles). It is named for German astronomer Johannes Kepler (1571–1630).

124. Grimaldi is a fantastic walled plain near the Moon's eastern limb. Its so large [222 km (38 miles) across], so relatively shallow [2.7 km (1.8 miles) deep] and so flat (its featureless floor is of smooth dark basalt) that it looks like a small sea. The smaller but impressively dark-floored **Riccioli** crater lies to its northeast and appears highly foreshortened due to libration effects;

it measures 146 km (90.5 miles) across and 2.3 km (1.4 miles) deep. Grimaldi is named for Italian astronomer and physicist Francesco Maria Grimaldi (1618–1663). Riccioli is named for Italian astronomer Giovanni Battista Riccioli (1598–1671).

125. Billy is a dark-floored crater 45 km (28 miles) across and 1.2 km (0.7 miles) deep. Its floor has been flooded by basalt almost to its rim. It makes a nice duet with equally sized **Hansteen** crater to the northeast; but Hansteen's floor has a higher reflective surface being marred by ridges, hills, and grooves. Billy is named for French mathematician Jacques de Billy (1602–1679); Hansteen is named for Norwegian astronomer and physicist Christopher Hansteen (1784–1873).

126. Letronne is a huge, lava-filled ruin of a massive impact crater in Oceanus Procellarum. It has all but been flooded by basalt; its northern rim is completely overrun by lava, forming an open bay. The crater measures 120 km (74 miles) across and a mere 1 km (0.6 miles) in depth. Its floor has some craterlets, but most interesting is a long north–south trending wrinkle ridge that arcs through the entire formation. It is named for Jean Antoine Letronne (1787–1848), a French archeologist.

127. Gassendi, the "twin" of Letronne (#126), is a phenomenal sight – one reminiscent of a phase in Letronne's life, in which the full crater walls were still visible. It lies on the northern border of Mare Humorum (T) and has a prominent dimple (**Gassendi A**) linked to its north, and the two seen together have been likened to an engagement ring. The floor is quite elaborate, with a complex central rise and a magnificent system of crisscrossing rilles – the most complex system on the visible side of the Moon. It has been the site of numerous transient lunar phenomena, many of which have been attributed to the possibility of outgassing from the rille system. Gassendi is named for French astronomer and mathematician Pierre Gassendi (1592–1655); Gassendi A was once known as **Clarkson**, in honor of British amateur astronomer and selenographer Roland Clarkson, but that name is not officially recognized.

128. Mersenius is a moderately sized impact crater west of Mare Humorum (T). It measures 84 km (52 miles) across and 2.3 km (1.4 miles) deep. If large-binocular observers can catch the light just right, they may see that the crater's floor is swollen, like a dome, rising some 450 meters (nearly 1,500 feet) above the surrounding floor. The swelling is believed to be the result of an upwelling of molten rock that never broke surface. The crater is named for French mathematician and physicist Marin Mersenne (1588–1648).

129. Cavendish is a moderately sized crater about two crater diameters southwest of Mersenius. It measures 56 km (35 miles) across and 2.4 km (1.5 miles) deep. It is named for British chemist and physicist Henry Cavendish (1731–1810).

130. Vieta is a moderately sized impact crater northwest of Shickard (#131). It measures 87 km (54 miles) across and a fantastic 4.5 km (2.8 miles) deep. The smaller crater **Fourier** lies to its southeast. It is named for French mathematician Francois Vieta (1540–1603).

131. Schickard is one of the Moon's largest walled plains. Located in the far southwestern highlands, it measures 227 km (140 miles) across and a shallow 1.5 km (0.9 miles) deep. The crater's walls are extremely eroded and its floor flat with variations in intensity – from dark patches to shadowed craterlets, to undulating rises. The lava-filled plain **Wargentin** lies to its southeast. Schickard is named for German astronomer and mathematician Wilhelm Schickard (1592–1635).

132. Mee is a wonderful crater 132 km (82 miles) across and 2.7 km (1.7 miles) deep and dates between 4.5 to 3.9 billion years ago. Its rim has been heavily worn and battered. It's named for the Scottish astronomer Arthur Butler Phillips Mee (1860–1926).

133. Schiller is yet another fantastic elliptical formation, no doubt the result of two joining impacts, with the connecting wall utterly destroyed. Being so close to the southern lunar limb, we see a highly foreshortened ellipse measuring 179 km (111 miles) × 71 km (44 miles); it is also an impressive 3.9 km (2.4 miles) deep. Its floor is smooth with some pits. The crater is named for German astronomer Julius Schiller (unknown–1627).

134. Phocylides is a large impact crater southeast of Schickard and is part of the Schickard chain: with ruined **Nasmyth** overlapping it to the north and the unusual walled plateau Wargentin to the northwest, which in turn, links to Schickard. Phocylides is named for Dutch astronomer Johannes Phocylides Holwarda (Jan Fokker) (1618–1651).

135. Bettinus is a moderately large crater near the Moon's southwest limb, northwest of the magnificent Bailly crater (#136). Bettinus measures 71 km (44 miles) across and 3.8 km (2.4 miles) deep. It is flanked by the similar-sized **Kircher** crater to the southeast and slightly smaller **Zucchius** crater to the northwest. Bettinus is named for Italian mathematician and astronomer Mario Bettinus (1582–1657).

136. Bailly, our last crater, turns out to be the most magnificent crater . . . at least in size. It measures 303 km (188 miles) across and 4.3 km (2.7 miles) deep, making it 47 km (29 miles) more extensive than Deslandres (#78). Unfortunately, Bailly is so close to the southwestern lunar limb that it is hard to observe unless there is an extreme libration. Thus, for the casual observer, centrally located Deslandres is the largest and most accessible crater on the near side of the Moon. Bailly's floor is complex and lava free, with dual impacts on its southeastern rim: **Bailly A** and **B**. Bailly is named for Jean S. Bailly (1736–1793), a French astronomer and orator.

CHAPTER 3

Eclipses: descents into darkness

Solar eclipses

I looked up and saw the Angel of Death sacrifice its soul to God in apocalyptic silence.

Of all the wonders in the heavens, a total eclipse of the Sun surpasses all in awe, beauty, and elegance. Totality can warp our perception of time, bring fear to the superstitious and joy to the enlightened. It can open the tap to our rawest emotions, and, for some, send the mind temporarily into another "dimension" or spiritual plane.

A total solar eclipse occurs when the Moon sails in front of our star and completely covers its luminous face. When it does, the Moon's shadow falls onto a tiny section of the Earth. The shadow's path is relatively small, measuring some 200 miles wide and covers only about 0.5 percent of Earth's surface. As the Moon moves in its orbit, and as the Earth rotates, the Moon's shadow moves across the face of the Earth along what's known as the *path of totality*.

Anyone fortunate enough to be swept into the shadow, will see (weather permitting) the dark face of the new Moon silhouetted against the diaphanous white wings of the Sun's outer atmosphere, or *corona*. No photograph can capture the individual experience of totality. The passage

---

WARNING!

* Failure to use proper methods when observing the Sun may result in permanent eye damage (retinal burns or thermal injury), which could cause or lead to blindness.
* Do *not* look at the Sun with your unaided eyes for any prolonged period of time.
* Do *not* use sunglasses, smoked glass, CDs, DVDs and CD-ROMs, film negatives, or polarizing filters, to look at the Sun; they are not safe for the purpose!
* Do *not* look at the Sun directly through binoculars without them being covered with specifically designed filters – filters that cover the objective lenses at the front end of the binoculars and let only about one part in 100,000 through, reflecting the rest.
* *Never* leave unattended mounted binoculars (without filters) pointed at the Sun, since even an instant's viewing could be devastating to someone's eyesight.
* Do *not* use solar filters at the eyepiece end of your binoculars. The focused light can burn through, or crack, the filters.
* Purchase solar filters only from reputable dealers.
* If in doubt about a filter's safety . . . *do not use it!* Do contact your local planetarium or astronomy club for more information.
* For more information, see page 5.

---

that opens this section, for instance, is how I felt when I saw totality for the first time in 1984. But my feelings have been different at other eclipses – sometimes I've been exalted, at other times more somber. The eclipse experience is never the same – which is why eclipse chasing has become such an attractive and popular sport.

Totality is a product of an intriguing historical "coincidence." Although the Sun is 400 times larger in diameter than the Moon, it is also 400 times farther away. So the two orbs look about the same size in our sky, which is roughly $\frac{1}{2}°$, or about half the diameter of your pinky's fingernail held at arm's length (when seen with one eye closed). This was not always the case.

Over the course of billions of years, tidal interactions between the Earth and Moon have caused changes in the

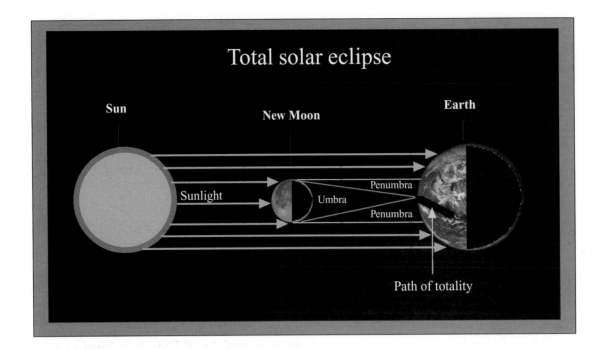

Total solar eclipse

Sun

New Moon

Earth

Sunlight

Umbra

Penumbra

Penumbra

Path of totality

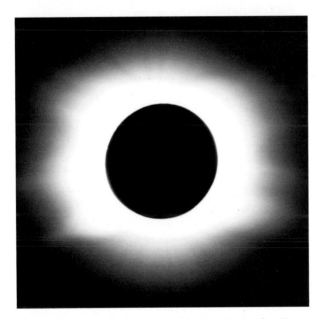

distance between these two worlds. Hundreds of millions of years ago, before humans appeared on Earth, the Moon was closer to our planet than it is today; close enough to have blocked the densest and brightest parts of the Sun's corona during a total eclipse. Now the Moon is slipping away from Earth at a rate of a few inches per year. Hundreds of millions of years in the future, the Moon will be too small to fully occult the Sun, so there will be no more total eclipses. We, then, are living during a privileged time in Earth's history when the Moon is the right size to fully block the Sun's face, offering us that rare glimpse of the solar corona, which is otherwise too dim to be seen in the daytime sky.

Myth and superstition

Totality has left its mark on humanity. The earliest account of a solar eclipse can be found in the *Shu Ching* – an ancient Chinese document most likely dating to the twenty-second century BC – which describes how "the Sun and Moon did not meet harmoniously." The document also tells us how the royal astronomers Hsi and Ho literally lost their heads because they failed to alert their emperor of the day's coming event.

In ancient China, eclipses were believed to be the cause of an invisible dragon eating away at the life-giving Sun. In ancient China, the name for "eclipse" was *chih*, which means "to eat." With enough warning, they believed, they could defeat the dragon by shooting arrows into the air, banging drums, and making other loud clanking noises during the eclipse. The practice never failed; the dragon always retreated, restoring peace and the light of day – thus strengthening their belief in the myth of man's power over this supernatural entity.

Many early cultures saw the eclipse as a dragon eating the Sun, including most of the people of northern Asia, Persia, Finland, and Lithuania. Other cultures envisioned a similar conflict; just replace the dragon with some other hungry beast. The Vikings, for instance, saw the "dragon" as a hungry wolf. In Africa (as in some parts of India), a snake did the devouring. The Hindus considered that it was a ravenous giant, while California's Serrano Indians imagined spirits of the dead gobbling up the Sun. And in Polynesia, a god that had been offended by neglect swallowed both the Sun and the Moon; appeasement came in the form of liberal presents and offerings until the god ejected the two luminaries from his stomach.

The nineteenth-century German story teller Jacob Grimm emphasized the impact of eclipses on "heathen" beliefs, saying the heathens associated them "with a destruction of all things and the end of the world." But while eclipses have brought about panic and fear through-out history, especially among the common masses, not all eclipses were ill omens (at least not to all involved) – a fact that gets overlooked in much of today's popular literature.

As Anton Pannekoek describes in his *A History of Astronomy* (Dover; New York, 1961), in the eighth century BC,

for instance, Babylonian astrologers not only predicted eclipses but foretold both good and evil events, such as this prophecy from Irasshi-ilu, the king's servant:

> On the 14th an eclipse will take place; it is evil for Elam and Amurru, lucky for the king, my lord; let the king, my lord, rest happy.

The fifth-century Greek historian Herodotus tells us a fantastic tale of how a total eclipse of the Sun, which had been predicted by Thales, the Milesian, stopped a war between the Lydians and Medes in Asia Minor [now Turkey; as told in *The Life of Cyrus* (The Religious Tract Society; London, instituted in 1799)]:

> A remarkable occurrence is mentioned by Herodotus as happening while Cyaxeres was extending his conquests towards Halys, being at war with the Lydians, a rival power in Asia Minor. During a battle between them and the Medes, the day was changed into nocturnal darkness, which put an end to the engagement.

If true, the eclipse led the two warring factions into peace.

Barbara Mann and Jerry Fields of Toledo University, Ohio, argue that a total solar eclipse in AD 1142 led the Seneca Indians of North America to adopt the Iroquois Great Law of Peace, making them the last of the five Indian nations to adopt this law offered by Deganawidah (the Peacemaker) and Aionwantha (Hiawatha). "The duration of darkness," Mann says, "would have been a dramatic three-and-a-half-minute interval, long enough to wait for

the sun; long enough to impress everyone with Deganawidah's power to call forth a sign in the sky."[1]

On the darker side of totality, the solar eclipse of June 7, 1415, helped to motivate a religious slaughter; on August 24, 1415, in "an aura of miracles and omens" (including the eclipse mentioned above), young Prince Henry of Portugal launched a crusade against a Muslim stronghold at Ceuta on the northwest African coast near the Straits of Gibraltar; the grim victory not only endorsed the Prince's strong belief in astrology but gave him a gory sense of triumph as he strolled the streets among the piles of enemy corpses.

Despite the marked increase in our understanding of eclipse science, time has not quashed the age-old superstitions associated with the visual phenomenon. During the January 22, 1898, total solar eclipse, which was visible, in part, over India, British Astronomical Association expedition members recorded the following scene:

> Throughout the dread two minutes of totality we were dimly conscious of an undertone of cries and wailing from the distant village . . . and the shout of relief and welcome with which the villagers greeted the return of sunlight was to us as loud and clear; and now that the crisis was past, the villagers gave themselves up to unrestrained rejoicing.

On October 24, 1995, nearly 100 years after the January, 1898 eclipse, I too was in India watching the Moon fully eclipse the Sun and recorded a similar experience. On that day, hundreds of spirited local villagers followed our eclipse caravan to a viewing site near Gavanoli. Shortly after sunrise, the Moon began to cover the Sun. The air was charged with chatter as the villagers watched us ready our equipment. But when daylight began to fade, and the air became filled with ghostly breezes, the villagers looked confused and apprehensive. Tension mounted until the moment of totality, when the crowd moaned and suddenly fell silent; many turned their backs to the Sun. In stark contrast, we, on the other hand, began to jump and shout with glee. When totality ended 53 seconds later, the tides suddenly turned; our moans of sadness were drowned out by an explosion of cheers that welled up from the villagers, who rushed at us with open arms and a garland of smiles – greeting us as if we were somehow responsible for the Sun's safe return.

Eclipse "families" and types

The January, 1898 eclipse is special to me because it belongs to a "family" of eclipses that includes the March 7, 1970 total solar eclipse – the event that gave me my first view of the Sun's corona, which I saw from my childhood home in Cambridge, Massachusetts.[2]

---

[1] From Bruce E. Johansen's 1995 article "Dating the Iroquois Confederacy," published in *Akwesane Notes New Series*, volume 1, nos. 3 and 4, pages 62–63.

[2] This eclipse was not total over Cambridge, Massachusetts, but see page 59 for an explanation of how I saw the Sun's corona.

The "family" is actually a cycle of eclipses known as a *saros*, from the ancient Sumerian/Babylonian word "SAR," which represents the quantity 3,600. Each saros cycle has a period of 18 years and $11\frac{1}{3}$ days. During that period, 41 solar eclipses will occur, with 10 of them being total. But here's the twist: while the eclipse paths of each succeeding saros are near repetitions of those of the preceding one, they all occur 120° farther west in longitude (thanks to the $\frac{1}{3}$ day in the saros cycle, which allows Earth to rotate a third of the way around its axis, or 120°).

So while totality occurred over parts of Massachusetts, on March 7, 1970, its sibling eclipse in the next saros cycle occurred on March 18, 1988 (exactly 18 years 11 days later); but the extra $\frac{1}{3}$ of a day caused the eclipse to occur 120° to the west over the west Pacific. The eclipses of the new series will also occur a little farther north if the Moon happens to be crossing the ecliptic from the north, or farther south if it is crossing the ecliptic from the south.

Before I discuss the drama of totality in more detail, you should know that not all solar eclipses are *total*. Sometimes the Moon covers only a part of the Sun, and we see a *partial* eclipse. At other times the Moon appears slightly smaller than the Sun's disk, and we see a *ring*, or *annular*, eclipse. And sometimes the Moon's disk so narrowly covers the Sun's face that we see a *hybrid* eclipse, meaning we see either a snug Moon–Sun-fit annular eclipse *or* a total eclipse depending on our location on Earth. To understand why we have this zoo of eclipses, we need to look at how the Moon orbits the Earth, as well as the Moon's shadow, in more detail.

All solar eclipses occur at new Moon, the time when the Moon in its orbit lies between the Earth and the Sun. But not all new Moons occur directly in front of the Sun. That's because the Moon's orbit around Earth is tilted 5° with respect to Earth's orbit around the Sun. Most of the

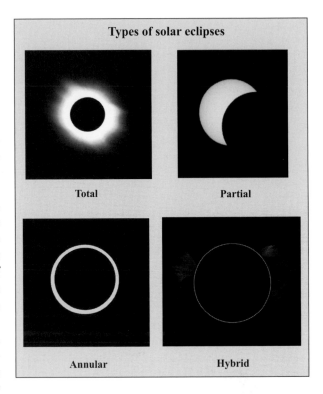

Types of solar eclipses

Total

Partial

Annular

Hybrid

time, the new Moon passes either north or south of the Sun's disk, and its shadow completely misses the Earth, so we see no eclipse; we also see no Moon, because the Sun is illuminating the side of the Moon facing away from us, leaving its near side in complete darkness, the lunar phase we call new Moon.

In the diagram below, note too that the Moon is not always the same distance away from the Earth. That's because the Moon orbits our world not in a circle but in an ellipse that carries it to within 221,000 miles at its closest approach (lunar *perigee*) or to 253,000 miles at its farthest point (lunar *apogee*). When the Moon is at or near lunar perigee, it appears slightly larger than the Sun in

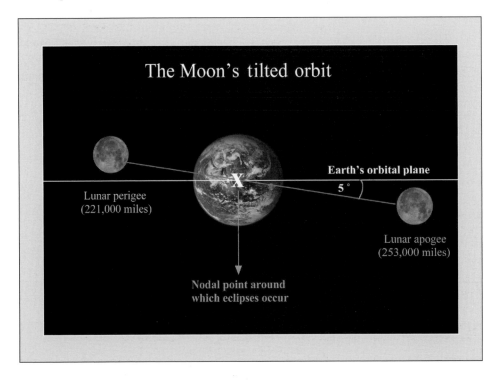

The Moon's tilted orbit

Earth's orbital plane

5°

Lunar perigee
(221,000 miles)

Lunar apogee
(253,000 miles)

Nodal point around
which eclipses occur

the sky. If the Moon passes centrally across the Sun at this time, we will see a total solar eclipse. When the Moon is at or near lunar apogee, it appears slightly smaller than the Sun in the sky. If the Moon passes centrally across the Sun's face at this time, we will see an annular event. The amount of Sun covered by the Moon, then, depends on the Moon's distance from Earth at the time of the eclipse.

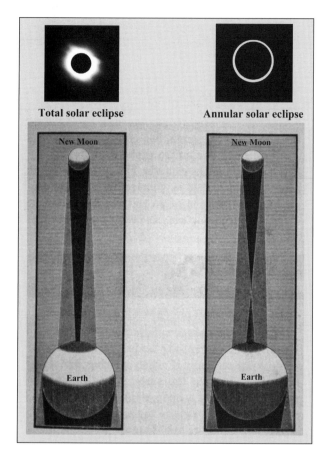

Total solar eclipse **Annular solar eclipse**

For a solar eclipse to occur, the Sun must be near one of the points where the Moon's tilted orbit intersects Earth's orbital plane (a *node*); the Moon also needs to be either ascending or descending past that node at new Moon. If it does, some part of the Moon's shadow will fall on the Earth. But the Moon's shadow, as with all shadows, is comprised of two parts, a dark inner portion called the *umbra*, and a fainter outer portion called the *penumbra*. The type of eclipse we see depends on which part, or parts, of the Moon's shadow hit the Earth.

If the Sun and Moon are not perfectly aligned – so that the Moon's dark umbral shadow completely misses the Earth, but parts of the penumbral shadow do hit it – a *partial* eclipse[3] occurs, where the Moon covers only a part of the Sun's face. All partial eclipses occur inside the penumbral shadow – which can cover large portions of the globe.

During a total solar eclipse, both the Moon's umbral and penumbral shadows are cast onto the Earth. Totality occurs inside the umbral shadow, which, over time, sweeps across the Earth's surface in a path. The cross section of the umbra has a maximum of 186 miles, but the projected width of the path on Earth's surface can be much wider. Those standing outside the *path of totality*, in the Moon's penumbral shadow, will see a partial eclipse. How much of the Sun's face is covered depends on the observer's location inside the shadow: the deeper one travels into the penumbra, the greater the partial eclipse. The deeper one travels into the umbra, the longer the total eclipse. The NASA eclipse map[4] on page 48 shows a section of the path of totality (the dark gray lines with circles) over Africa for the total solar eclipse of March 29, 2006. The lines parallel to it show the penumbral shadow and the percent of the Sun's disk which is eclipsed by it.

All eclipses begin with *first contact* (when the Moon's advancing limb first touches the Sun) and end with *fourth contact* (when the Moon's trailing limb last touches the Sun's outside limb). And these are the only contact points that occur during a partial eclipse. Total and annular eclipses have second and third contacts (see the bottom diagram on page 48).

During total eclipses *second contact* occurs when the Moon's advancing limb first touches the inside edge of the Sun's limb (this marks the beginning of totality). *Third contact* occurs when the Moon's trailing limb last touches the Sun's inside limb (this marks the end of totality). During annular eclipses, second contact occurs when the Moon's trailing limb first touches the inside edge of the Sun's limb (this marks the beginning

---

[3] Of the maximum number of solar eclipses that can occur per year (namely, five), four of them will be *partial*. Partial eclipses are, in fact, the most frequent of all solar eclipses; in the 5,000-year period beginning in 2000 BC, 35 percent of all eclipses have been partial.

[4] Map courtesy of Fred Espenak, NASA/Goddard Space Flight Center.

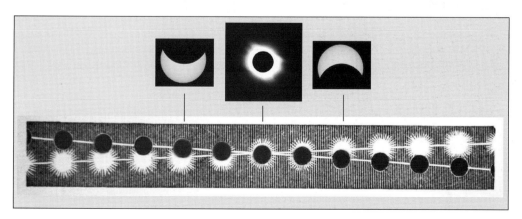

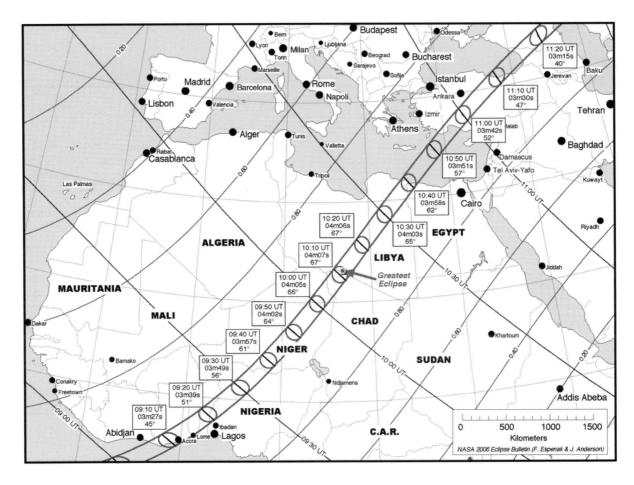

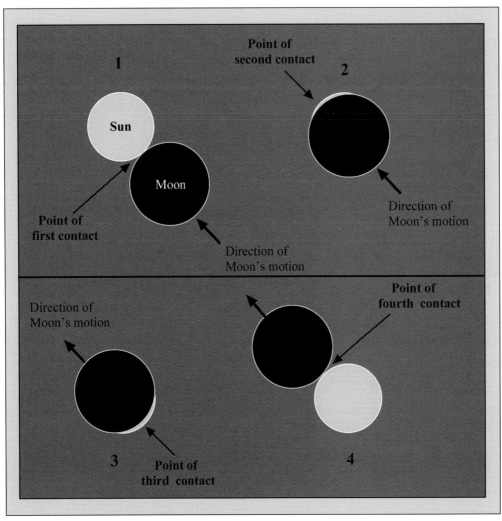

of annularity); third contact occurs when the Moon's advancing limb last touches the Sun's inside limb (this marks the end of annularity).

Since portions of the Sun remain visible during a partial eclipse, or the partial phases of a total solar eclipse, you will need to protect your eyes by either projecting the Sun's image with your binoculars or covering the binocular's front lenses with a safe solar filter (see page 5).

Let's look now at some of the wonders that keen observers can experience during the partial stages of an eclipse. The phenomena are the same for either a partial eclipse of a certain magnitude or the partial phase of a total eclipse of that same magnitude. In other words, it doesn't matter whether you're experiencing a maximum partial eclipse of 75 percent, or if the Sun is 75-percent eclipsed during a total solar eclipse, the related phenomena are the same. Mine is a rough guideline not a dictum. When it comes to the timing of solar-eclipse events, nothing is written in stone. Expect the unexpected.

The partial phases and their effects

WARNING!

* **Failure to use proper methods when observing the Sun may result in permanent eye damage (retinal burns or thermal injury), which could cause or lead to blindness.**
* **Do *not* look at the Sun with your unaided eyes for any prolonged period of time.**
* **Do *not* use sunglasses, smoked glass, CDs, DVDs, and CD-ROMs, film negatives, or polarizing filters, to look at the Sun; they are not safe for the purpose!**
* **Do *not* look at the Sun directly through binoculars without them being covered with specifically designed filters – filters that cover the objective lenses at the front end of the binoculars and let only about one part in 100,000 through, reflecting the rest.**
* ***Never* leave unattended mounted binoculars (without filters) pointed at the Sun, since even an instant's viewing could be devastating to someone's eyesight.**
* **Do *not* use solar filters at the eyepiece end of your binoculars. The focused light can burn through, or crack, the filters.**
* **Purchase solar filters only from reputable dealers.**
* **If in doubt about a filter's safety . . . do *not* use it! Do contact your local planetarium or astronomy club for more information.**
* **For more information, see page 5.**

One of the most exciting moments of any eclipse is the detection of *first contact* – when the Moon's advancing limb kisses the Sun's shoulder before taking its first bite. This sensual moment occurs precisely on prediction and is a shining example of science's triumph over the age-old belief that eclipses were supernatural events warning man of his follies on Earth.

Fully aware of the impact of eclipse predictions on humanity, Pliny the Elder (AD 23–79) praised, in his *Natural History*, those great men of science who discovered the laws that govern the motion of the Sun and Moon, who "freed the miserable mind of man" from fear of eclipses. "Praise be to your intellect, you interpreters of the heavens, you who comprehend the universe, discoverers of a theory by which you have bound gods and men!" Pliny extolled.

Actually, the precise moment *you* see first contact may be delayed by a few seconds, especially if you are uncertain as to exactly where along the Sun's limb this event is scheduled to occur.[5] The anticipation and uncertainty adds an element of surprise and fun to the task – especially if you're part of a group of eclipse enthusiasts all vying to be the first to spy it. The victor usually gives a shout, which can trigger an avalanche of excitement as more and more people catch sight of the Moon's dark limb slipping neatly into place.

The greater the aperture and magnification, the better your chances of catching first contact closer to the predicted time: telescope users have an advantage over binocular users; those with large binoculars have the edge over others with smaller ones; and all binocular users have the upper hand on those using their unaided eyes and a safe filter. During the June 21, 2001 eclipse, I detected first contact one second after the predicted time using binoculars; but the black notch of the Moon did not become apparent to the unaided eye (as seen through a safe #14 welder's glass) until a couple of seconds later. Today, we know much more about the orbital dynamics of the Moon, as well as the physical size of the Sun and other determining factors, so first-contact timings, especially with digital imaging and other recording devices, are much more accurate.[6]

Brightness-contrast illusions

As the eclipse progresses and the Moon covers more and more of the Sun's face, pay close attention to the advancing limb of the Moon. If you stare just right, you

---

[5] While newspapers, television, and astronomy magazines may give you eclipse contact times for major cities, the premier site for learning about the visibility of eclipses is the NASA Eclipse website (http://eclipse.gsfc.nasa.gov/eclipse.html) operated by "Mr. Eclipse" himself, Fred Espenak of the NASA/Goddard Space Flight Center. Software programs, like Eclipse Timer (http://www.eclipsetimer.com/index.htm), are also available.

[6] *Visual* timings of fourth contact are much more precise than those of first contact because observers can follow the Moon's receding silhouette to the exact point of exit on the Sun's limb.

may see a luminous band surrounding the visible portion of the Moon's limb. Some nineteenth-century observers ventured to guess that this bright border was due to the refraction of sunlight through a lunar atmosphere; the evidence seemed all the more convincing, since the luminous band was also photographed at several different eclipses.

But other astronomers, such as the Rev. James Challis, a Fellow of the Royal Society, argued in 1864 that it was purely a physiological effect created by the eye's response to seeing a "common boundary of a light and a black space under circumstances of indistinct vision . . . " A year earlier, George Biddell Airy, England's Astronomer Royal, had already concluded that it is "strictly an ocular nervous phenomenon, not properly subjective but sensational – a mere effect of contrast."

Indeed, we know today that the luminous band preceding the Moon's limb is a brightness-contrast illusion. If you stare at a black object long enough against a bright background, a portion of the eye's retina will become fatigued. Look at the illustration below. If you stare at the edge of the black crescent for 10 to 20 seconds, a negative (or bright) afterimage of the dark object will form. Small, natural eye movements (called saccades) will carry the bright afterimage beyond the borders of the black crescent, creating the illusion of a luminous band collaring the Moon's silhouette. Anyone with normal vision can see it.

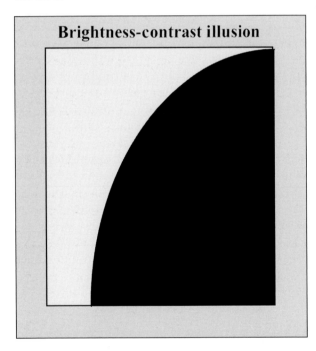

Brightness-contrast illusion

If the Sun's face is marred by sunspots, look for a similar illusion – one appearing as a luminous ring around the black spot. What happens when the dark limb of the Moon, accompanied by the brightness-contrast illusion, approaches the spot? Can you see the two illusions simultaneously, or can you see only one or the other, depending on your eye's point of focus? Do the results vary with the size of the spot? Do they vary with each eclipse? There's much to record in your binocular observations.

Illusion or not, seeing the Moon's limb cover a sunspot is exciting. The event breaks up the otherwise monotonous passage of the Moon across the Sun's face. Unlike the crisp, snapping out of light during a lunar occultation of a nighttime star, it can take the Moon some time to cover a sunspot, depending on the size of the spot. If your binoculars have enough magnification, it is possible to see the spot briefly "attached" to the Moon's limb, giving rise to a temporary illusion that a black mountain is projecting from the lunar surface. If the atmosphere is at all unsteady, the spot and limb may perform a celestial swing dance – one in which the quivering spot jumps toward and away from the Moon's body in uncontrollable jerks until the two fully unite.

Crescent shadows

Another fun way to pass the time during the partial phases is to stand beneath a tree and look at its shadow on the ground for crescent images of the partially eclipsed Sun. These are projections of sunlight passing through tiny, pinhole-like spaces between leaves. Under excellent circumstances, the tree's shadow can be dappled with hundreds upon hundreds of these crescents (some large, some small), which flit about in the breeze like winged fairies in flight.

One of my fondest childhood memories is of a partial solar eclipse that I watched on my parent's bedroom floor on the afternoon of July 20, 1963, at the age of six. At the time we were living on the top floor of a triplex on Holworthy Street in Cambridge, Massachusetts. My parents' bedroom window faced west, allowing us an unobstructed view of the eclipse, which started nearly half way up the sky. But I didn't look directly at the Sun's disk. Instead, my mother let the slanting rays of the eclipsed Sun pass through the small rope holes in the window's Venetian blinds, which she had lowered for the event. And *voila*, as if by magic, a neat row of solar crescents appeared on the bedroom floor, offering mom and I a grand way to watch the eclipse in complete safety from the comfort of our home.

Visual effects on the environment

If the Moon is to cover more than 50 percent of the Sun's face, periodically record changes to the environment after first contact. Note the color and quality of light on and around some nearby objects, then do the same for the distant landscape. Also record the color and conditions of the sky, especially around the horizon, and journal the behavior and sounds of any wildlife. In other words, open all your senses at the start of the eclipse, and keep them on high alert as the Moon blocks more and more of the Sun's light. Almost everything you see will vary from eclipse to eclipse, so be sure to keep good records.

Air temperature drops noticeably when the Sun is about 50 percent eclipsed. The dip is usually accompanied by caressing breezes that can feel like the tender brush of a lover's fingers against your skin. The quality of light also changes, turning a muted shade of yellow, as if the Sun had slipped into a blanket of smoke. Earth tones seem to tarnish and reflect an ashen hue. Details sharpen and have more contrast. Your eyes should be able to penetrate more deeply into shadows.

If you project the shadow of your finger onto a bright sheet of paper, you'll see that one side of the finger appears fuzzier than the other. This enhanced fuzziness appears on the side of the Sun being eclipsed. In fact, the farther you move your finger away from the sheet, the more curved it will appear!

When the Sun is three-quarters covered, the colors of the Earth and sky begin to fade like an old print. After that, changes occur ever more rapidly as the Sun becomes a thinner and thinner crescent. The following account illustrates how the partial phases leading up to a total eclipse can affect the environment. It's based on my journal notes from the February 26, 1998, eclipse that my wife, Donna, and I observed from a yacht near Pinta Island in the Galapagos:

75 percent eclipsed. The western sky is yellow; the eastern sky is slate. Bright areas on the boat lost their luster. The faces of the people surrounding me turned steely blue, as if some alchemist had turned their skin to metal. The wind gusted.

90 percent eclipsed. The temperature in the Sun plunged to a low of 88 °F - a dip of 27 °F since the start of the eclipse. Chocolate, which had melted in the Sun, began to harden. Frigate birds began to circle high overhead, while seagulls flew abreast of us, looking as if they wanted to roost.

95 percent eclipsed. Donna saw the water's surface being pelted by an invisible rain. Thousands of tiny circles expanded in the water, snapping and popping as they appeared. Yet the sky was blue and we couldn't feel a drop of rain. Then Donna discovered the cause: night fish had surfaced and were snapping away in a feeding frenzy.

98 percent eclipsed. The western horizon looked bruised and ugly, apocalyptic. Clouds turned pink, the ocean deep blue. Wedges of tangerine light slipped into the moody northern and southern skies. Venus and Jupiter snapped into view. All around me light drained away. I felt as if I was stepping into a dream where people stand like ghosts in a faded reality.

Shadow bands

About three minutes before totality, when the Sun is a narrow crescent, periodically draw your attention away from the Sun and look around with your unaided eyes for *shadow bands* – a series of alternating dark and light lines that wash across the ground, walls, and other structures, like the moving waves of light and shadow reflected from the surface of a rippling pool. Many eclipse watchers prepare for this dim and elusive atmospheric effect by placing a large white sheet by their feet, so all they have to do is look down at the appropriate moment to see them. Some observers also mark the points of the compass on or around the sheet to determine direction of motion.

Shadow bands have no doubt been seen since man first began to fathom eclipses. But the first vivid account of them was made by German astronomer and painter Hermann Meyer Solomon Goldschmidt in 1820 during an annular solar eclipse (see page 60). Shadow bands have been observed at most eclipses ever since.

When George Biddell Airy noticed this "strange fluctuation of light" during his first total solar eclipse in 1842, he said the effect was "so striking that in some places children ran after it and tried to catch it with their hands." And while many observers have reported seeing them shortly before and after totality, some nineteenth-century observers claimed to have seen them during totality – a task no modern visual observer has attempted, at least to my knowledge!

At first, some astronomers theorized that the bands were diffraction fringes bordering the Moon's shadow. Around the turn of the twentieth century, their thinking began to focus on the irregular refraction of light coming from the Sun's narrow crescent. But these early attempts to decipher the shadow-band code were continually thwarted by the highly irregular nature of the sweeping bands.

Shadow bands are indeed highly variable in their appearance and have the most mystifying behavior of all the eclipse phenomena; that's why they're deserving of your attention. For the 2006 total solar eclipse, Andrew Greenwood, chairman of the Macclesfield Astronomical Society in Britain, observed from the "very alien" Libyan Sahara, about 50 miles south of the oasis town of Jalu. There he saw a prominent and lengthy display of shadow bands 43 seconds before second contact. The bands appeared "very contrasty and fairly hard-edged," being 1 to 1.5 inches in diameter and spaced about 3 inches apart. "You could almost feel them!" he says. "It was very much like being underwater in a pool with the refraction from the ripples above."

Modern studies have shown shadow bands to be an atmospheric phenomenon – one that arises as light from the Sun's narrowing crescent interacts with turbulent bundles of air in Earth's atmosphere. Each bundle acts like a lens, which can either separate the incoming beam into small twinkling packets or split it into long waves that interfere with one another – constructively and destructively – which we see as an interference pattern of long light and dark bands. The bands can also break up into individual packets in response to shifting air currents, which can shuffle or swirl the atmospheric lenses

about to create the myriad lighting effects we see on the ground.

The photo illustration above shows the enhanced appearance of these mysterious serpentine shadows. The sinuous patterns move across the ground with typical speeds of a few yards per second in the direction perpendicular to their length. As totality approaches and the crescent narrows, the spacing between the light bands decreases while the contrast between the light and dark bands increases. It is no surprise, then, that one of the reasons why so many people have never seen shadow bands is that they appear most magnificent at a time when most observers have their eyes glued on the Sun in expectation of the stunning phenomena immediately preceding totality!

Diamonds, rubies, necklaces, and rings

About two minutes before totality, block the Sun's remaining crescent with a nearby foreground object, remove the filters from your binoculars, and carefully look for the trailing (western) side of the Moon through your binoculars; its rounded form should be weakly silhouetted against the Sun's feathery inner corona. [**Do not look at the uneclipsed Sun, no matter how small the crescent, directly through binoculars! Failure to do so may result in permanent eye damage (retinal burns or thermal injury), which could cause or lead to blindness**.] The appearance is that of a smoldering tiffany *ring* with a somber interior; this aspect actually occurs much sooner (see page 59), but it is now most apparent, especially through binoculars.

And what a sight! The appearance of the early corona is something greatly understated and highly neglected, yet it offers binocular viewers a stunning view of what I call *coronal hair*, as well as some of the most pure and pastel colors seen in the heavens.

Coronal hairs are fine filamentary structures resembling countless fibers radiating from the Moon's limb like the silken threads of milkweed seeds blowing in the breeze. The hairs comprise the inner fabric of the larger-scale coronal *petals* and *streamers*, which dominate views during totality. All these phenomena are plasma flows associated with solar prominences and other eruptions from the Sun's surface that eject streams of particles into space, where they are dispersed by the solar wind.

The corona's color is even more mysterious and, apparently, highly subjective. Few modern descriptions exist of the corona's color before second contact (or after third contact) and it may be one of the most understudied of all visual eclipse phenomena. Part of the problem is that the colors are subtle – shaded breaths of light – and the eye is still adjusting to the growing darkness. But binoculars help to enhance the view. Seeing the color also requires the use of direct vision, not the averted vision that visual astronomers so much rely on for observing faint objects at night. You want light from the corona to strike the eye directly, so that it washes over the color-sensitive cone cells that line the eye's central fovea. The most glorious color I've ever seen was virgin teal – a color so pure it was hypnotic.

Seeing the corona's color at the magical moments just before and after totality is most important, because, once totality starts, the corona's brightness increases dramatically, which tends to wash out any subtle hues. Indeed, most observers see the corona shining white during totality. Nevertheless any colors you might see shortly before or after totality will probably depend on many factors, including the color sensitivity of one's eyes and the quality and conditions of the air, which can vary from location to location.

Modern measurements of the corona's true color reveal it to be white, with a touch of green. But that doesn't mean the various colors observed in the corona are illusory. The blue shades we see from Earth may actually be coronal light (which has the same brightness of the full Moon) being scattered by the molecules that comprise our atmosphere; the night sky, by the way, does look blue under full moonlight. Add a touch of dust or pollutants to the air and the corona may become more aquamarine.

Under very dusty conditions, or perhaps during years when volcanic aerosols are present in Earth's stratosphere, the corona may take on warmer shades. Indeed, during the 2006 eclipse, which I observed from the dusty Libyan Desert in Egypt, the corona's color changed dramatically between second and third contacts. In the minute prior to second contact, the inner corona as seen through 10 × 50 binoculars was a frosty blue, like pale shadows cast onto snow. During totality the polar brushes were white trimmed with green. And in the minute following third contact, the frosty blue inner corona was fringed with red. When sunlight returned I noticed dust blowing in the air. Coincidence?

How does interplanetary dust affect our view of the corona? If a sungrazing comet passes by the Sun shortly before an eclipse, and deposits additional dust to the interplanetary medium, will the corona redden? As always, never shy away from recording exactly what you see. Every experience is a key that may unlock a door to understanding.

About 30 seconds before totality, quickly place the filters back on your binoculars and immediately turn your attention to the thinning solar crescent. Some 15 seconds before second contact, the Moon's advancing limb will grow stubby black fingers that stretch radially across the remaining sliver of sunlight – joining the lunar and solar limbs and splitting the Sun like quicksilver into a number of quivering drops of liquid sunshine. Commonly known as *Baily's Beads*, these drops are named after Francis Baily (1774–1844), the English astronomer who greatly promoted their appearance in 1836, noting that they were the result of sunlight passing through the occasional depression, or valley, along the Moon's rough and ragged limb.

Beads

But this moniker is unfortunate, since Baily was by no means the first person to detect, nor communicate, the

phenomena . . . not even in 1836! The earliest reference may, actually, be described metaphorically in the Book of Revelation (Chapter 12, verse 1): "And there appeared a great wonder in heaven; a woman clothed with the sun, and the moon under her feet, and upon her head a crown of twelve stars." Edmond Halley (1656–1742) first observed the phenomenon during the total solar eclipse of May 3, 1715[7] and published his observations in Volume 29 (1714–1716) of the *Philosophical Transactions of the Royal Society*.

Rather than name the beads after any one individual, I prefer to call the phenomenon the diamond necklace, after the many reports of the phenomenon in the nineteenth century that gave notice not only to the "beads" but also to the black interstices connecting them. (Interestingly, many modern observers fail to notice anything but the beads, while eighteenth- and nineteenth-century observers found the black spaces between them of equal interest.) Some early accounts liken the phenomena to "a luminous string of beads." Others saw "irregularly shaped fragments which at no time resembled beads." Still others saw "diamond-shaped beads, which appeared to adhere to each other and to the limbs of the sun and moon."

Indeed, more than a string of beads, I liken them to a jeweled necklace, though I too was not the first to do so. At the July 28, 1851, total solar eclipse, for instance, Mr. Dunkin, one of the assistants at the Royal Observatory who observed totality from Christiana, a Greek island in the Cyclades, logged the following account:

> About fifteen seconds before the beginning of total darkness, the narrow line of sun broke up into numerous small particles or beads of light. They were of different sizes, some being merely points, while others appeared more elongated; their appearance was of intense brilliancy, and the only thing they can be compared with is a necklace of diamonds.

At that same eclipse, the famous "eagle-eyed" English amateur astronomer William Rutter Dawes (1799–1868), who observed totality from near Engelholm, in Sweden, noted what I find is *the* most interesting aspect of the diamond-necklace phenomenon:

> I was particularly struck with the fact that the lunar mountains broke through the crescent while it yet appeared to be considerably broader than the extent of their own projection beyond the general outline of the moon's circumference. This was observable in the case of the largest of them, which divided the crescent so long before it seemed capable of doing so, as to take me quite by surprise. The lunar mountains had thus the appearance of being drawn out into narrow black threads, reaching to the exterior edge of the Sun. This curious phenomenon I attributed, on subsequent reflection, to the effect of irradiation,

---

[7] In fairness, it should be noted that the famous comet that bears Halley's name was not discovered by him. Halley, however, was the first to compute the comet's orbit and predict its return in 1758.

increasing the apparent length of the prominences which broke through it.

In other words, each lunar mountain – especially the greatest ones – were subjected to the famous black-drop effect – an illusion commonly witnessed during solar transits of Mercury and Venus (see page 83). Beyond irradiation, the explanation for the effect (at least those affecting binocular users) include the smearing effects of atmospheric seeing, one's eyesight, the quality of the binocular optics, and the magnifying power of the binoculars.

In the final seconds before totality, the smallest beads at the extreme ends of the solar crescent start to evaporate like snow under a hot Sun. The length of the diamond necklace rapidly shrinks. In its final stages, some of the beads merge like drops of water, until the necklace morphs into one large and magnificent *diamond* that sits atop the gossamer coronal ring; this is the famous *diamond ring* that heralds the start of totality. As soon as the diamond ring appears, you must must quickly *remove your safe viewing filters* and get ready for the miracle of totality.

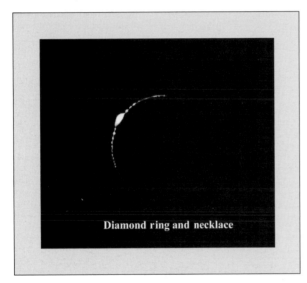

Diamond ring and necklace

If you are fortunate enough to look up at the sky at the right moment, just before the Sun totally winks out, you may see with the unaided eye the *Moon's shadow* rush up from the western horizon with supersonic speed and *SLAM* into the Sun at the very moment of second contact. The collision unleashes a sudden and breathtaking explosion of coronal light, whose pearly luster splashes out from a jet black Moon fringed with solar blood, squirting in places as if from severed arteries.

The blood-red fringe is the Sun's chromosphere, which may, at times, be seen as rubies strung out along the diamond necklace in the final seconds before totality (see the photo illustration at top right). The chromosphere remains visible beyond the Moon's eastern limb for only a few brief seconds, so be sure to have your *unfiltered* binoculars ready to catch this fleeting appearance of the Sun's middle atmosphere. Exceedingly bright compared to the pale corona, the chromosphere looks like a hedgerow of molten carmine flames.

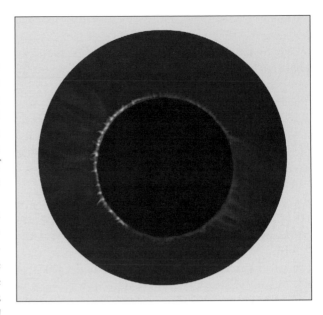

More impressive are the "squirts" of blood, which are actually gaseous *prominences* erupting into the corona from the solar surface. Photographs, as well as views through hydrogen-alpha filters, show their color to be deep red, but this is far from reality. Andrew Chaikin, author of the 1994 book, *A Man on the Moon* (Viking/Penguin Books; New York), told me that he likened their color during the 1991 eclipse to that of a magenta road flare that's just been ignited. Prominences really are hot magenta – a lovely mixture of yellow helium emission and red, blue, and violet hydrogen emissions. But other conditions, such as the state of Earth's atmosphere, the temperature of the prominence, its distance from the solar limb, may contribute to what color we see. For instance, at the 2006 eclipse in Egypt (with dust in the air), the prominences appeared as a very pale salmon pink.

Shaped by the Sun's powerful magnetic field, prominences can extend over thousands to hundreds of thousands of miles and appear as magnificent arcs, loops, sprays, or filaments; the largest can be spied during the diamond-ring stage and remain in view longer than chromospheric hedgerows, which themselves are the smaller, short-lived eruptions called *spicules*. If you're lucky, you may also catch sight of some arching filaments in the process of falling back to the solar surface as a molten rain.

In the mid-nineteenth century, some observers saw the chromospheric hedgerows as "resembling in outline the tops of a very irregular range of hills." Dawes and Airy referred to them as *sierras*, and proposed the existence of mountains on the Sun. Perhaps with fiery mountains in mind, Charles Babbage of Manchester Square proposed in 1851 that the "pink prominences" were a "species of volcanic action" connected with sunspots. "Those who have observed the dense lofty column of smoke arising in a clear calm day to great heights above a crater," he explains, "have also occasionally observed the top of a column of smoke, on reaching a gentle current of air, pass on horizontally to great distances. Such a horizontal column, seen endways from a point at great distance from our planet, might appear to be entirely disconnected from it by any continuous line of smoke."

Once the moving Moon clears the corona of any chromospheric "smoke," lower your binoculars for a moment and take in the glory of totality with your unaided eyes. Few are ever disappointed.

Totality[8]

If you ever want to taste universal peace and harmony, experience the joy of total spiritual freedom, or feel your soul washed clean of any ill, then you must see a *total* solar eclipse. A partial eclipse of 99 percent, or even the last seconds before totality, exciting and dramatic as they are, do little to measure up to totality. I have been to funerals, and I have watched bodies being lowered into the grave, but neither of these events, I imagine, will compare to experiencing death itself. Yet inside the path of totality, we all die, if for a moment. We're buried alive in the "inky" soil of the Moon's shadow. Looking up from our well of darkness, we see the awful Moon hanging before us like a dark angel with wings of light. A second of totality can give one centuries of understanding as to how man could put his faith in God. As reported in an 1860 *Monthly Notices of the Royal Astromical Society*, consider the experience of Lieutenant James Melville Gilliss at the

September 7, 1858, total solar eclipse, which he observed near Olmos, Peru:

> . . . the scene thrilled me with excitement and humble reverence. Nor was it less effective upon others. Two citizens of Olmos stood within a few feet of me, watching in silence, and with anxious countenances, the rapid and fearful decrease of light. They were wholly ignorant that any sudden effect would follow the total obscuration of the sun. At that instant, one exclaimed in terror, "*La Gloria!*" and both, I believe, fell to their knees, filled with awe. They appreciated the resemblance of the corona to the halos with which the old masters have encircled their ideals of the heads of our Saviour and the Madonna, and devoutly regarded this as a manifestation of the Divine presence.

The majesty of solar eclipses may be symbolized on the lintels of early Egyptian temples dating to about 3,000 BC, where we see carved representations of the winged globe, or winged solar disk. As author David Todd explains in his 1922 book, *The Story of the Starry Universe* (P. F. Collier & Son Company; New York), there's a "bare possibility" that the wings were suggested by a "type of the solar corona as glimpsed by the ancients." If true, the winged Sun may represent Horus of Behdet, who flies daily across the sky, doing battle with the evil Set – a dark god associated with terrifying events, such as eclipses.

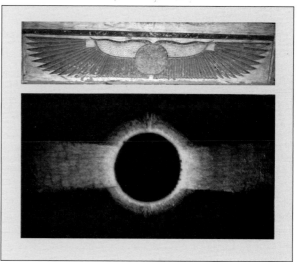

The ancient symbolism of light (good) battling darkness (evil) remains with us to this day. Many of our fairy tales, books, and movies – like those written by horror mogul, Stephen King, whose plots deal specifically with the conflict between good and evil – owe their origins, in part, to total solar eclipses. Perhaps, then, when we look up at the totally eclipsed Sun, we see a reflection of our own soul, just as Oscar Wilde's fictitious character, Dorian Gray, did, when he looked upon the possessed portrait of himself. As Gray realized, "Each of us has heaven and hell in him."

It would be foolish of me to tell you how you should feel during totality, which is a highly personal experience.

---

[8] Note that everything leading up to totality happens in reverse after totality, so I will not repeat the observations that can be made after third contact.

No, the emotional impact of totality is best left for you to explore. What I can do is introduce you to the many wonders visible to the unaided eyes and binoculars during totality. But be aware that the visibility and appearance of these phenomena vary from eclipse to eclipse.

One thing you should know up front is that during the total phase, the sky does not turn as dark as night. The actual sky brightness depends on a number of factors, including the intensity of the Sun's corona (which shines about as bright as the full Moon), the amount of horizon light diffusing into the shadow, and the albedo (reflectivity) of your immediate surroundings. Generally speaking, the sky gets about as dark as it does roughly 45 minutes after sunset in the spring and fall. Totality will appear darker if you are standing on a grassy knoll as opposed to a snowy field.

Bright stars and planets will appear during totality, and though it is a novelty to see them in the day, I would not recommend spending much time scouting them out; doing so takes precious time away from the rarer eclipse phenomena. That said, if Mercury is well placed, do look at it. We can see every other planet, except Mercury, in a dark sky, so you may be startled by its appearance. Usually Mercury appears as a "tiny" rosebud of light in the twilight sky, which sets not long after the Sun (see page 74). But how shockingly bright and white it appears during totality, especially if the Sun is high in the sky, so that it is away from the dense atmospheric contaminants that redden its disk near the horizon.

The corona's structure and color

The most important feature to focus on is the Sun's corona – something that you will most likely never see again unless you are fortunate enough to see another total solar eclipse.[9] In fact, one of the reasons people enjoy traveling to eclipses is to see the phenomena they missed during the previous totality. And though the longest possible totality is 7 minutes and 31 seconds, chances are you'll have only a few precious minutes in the Moon's shadow; and even that will pass all too swiftly. As the adage goes, totality, no matter how long its duration, seems to last only 8 seconds.

The structure of the Sun's white-light corona varies from eclipse to eclipse. What shape you see depends largely on the configuration of the Sun's global magnetic field, which changes during the 11-year sunspot cycle. During periods of solar maximum, coronal petals (tipped by long steamers) fully surround the Moon's silhouette, so it looks like a snow-covered rose in bloom. At times

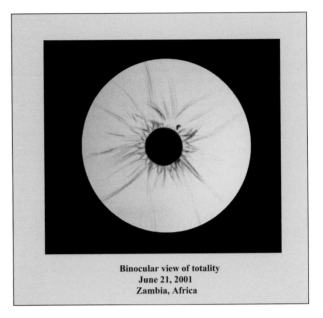

Binocular view of totality
June 21, 2001
Zambia, Africa

of solar minimum, the coronal petals stretch out majestically mainly along the Sun's equatorial plane; but not at its poles; that's because during solar minimum, the magnetic field lines at the Sun's poles are open, allowing gases to escape freely into space rather than looping back to the surface, as they do at the equator. The difference

Solar maximum corona

Solar minimum corona

[9] A total solar eclipse is an extreme rarity for those who do not like to travel. For any given town on Earth, a total solar eclipse will occur there, on average, only about once every **375** years! A total solar eclipse occurs some place on Earth about every $1\frac{1}{2}$ years. Fortunately, eclipse tourism is a thriving industry and affords many people not only a chance to see a total solar eclipse but also many interesting parts of the world.

between the solar maximum and minimum coronas is most pronounced through binoculars, which allows you to see the finer structural details in these coronal loops. At solar minimum, however, the Sun's poles have an equally fascinating, though dimmer, coronal phenomenon: the delicate polar brushes – feathery rays that rise from the Moon's cold black disk-like pale streaks of wind-blown snow or ghostly fingers of dawn.

If you lower your binoculars and sweep your eyes across the corona, you may see its large-scale petals set against more gossamer structures of dim coronal light. During the 1998 eclipse over the Galapagos (one year after solar minimum), I saw the corona's long equatorial petals and streamers against a fantastic halo of dim coronal light several degrees in extent. Contrarily, during the June, 2001 eclipse, which I saw from Zambia (one year after solar maximum), the Sun's symmetrical ring of coronal petals was centered on a 10°-wide equatorial band of faint coronal light (see the drawing below). In the former case, I wondered if the Sun's corona was not shining through a layer of high ice crystals in our upper atmosphere. In the latter case, I could not help but notice that the dim band was in the plane of the ecliptic (note the position of the planet Jupiter near the equatorial band of light) and wondered if it wasn't the corona extending into the zodiacal light – the milky band of dust (from comet tails and asteroid collisions) lying in the plane of the inner solar system out to about the orbit of Jupiter.

As I've mentioned, the color of the corona is white with a touch of green. The famous astronomical artist, Étienne Léopold Trouvelot, made a most compelling observation of this color during the July 29, 1878 total solar eclipse, which he observed from Oregon, in the Wyoming Territory. "On all sides of the sun's hidden disk," he noted, "the *corona* shows its pale greenish light extending in halo-like rays and streamers." He later published a remarkable colored drawing of this phenomenon.

At the December 4, 2002 total solar eclipse, which Donna and I observed aboard the *Marco Polo* in the Indian Ocean, we also saw the Sun's corona shining green. Shortly after second contact, the view through 10 × 50 binoculars showed the dark lunar limb surrounded by a slender ring of powder-blue coronal light, which itself was bounded by a luminous flower of pale, lichen-green,

coronal petals; these colors were magnified by the striking contrast of a ruby necklace adorning the Moon's western limb. Admittedly, we saw the eclipse through some high cloud, and the colors might have been due to diffraction effects. But after giving the matter thought, I believe that the clouds merely acted as a neutral density filter, cutting down the contrast between the dark Moon and the bright corona, allowing the corona's true and subtle colors to shine forth.

Subtle color-contrast effects may play a role in what color the corona appears to the unaided eye. If one looks at the marigold light that rings the horizon, outside the Moon's umbral shadow, then returns his or her gaze to the pearly corona, a bluish afterimage can wash over the corona. And, as I have mentioned before, contaminants in the atmosphere may redden the corona's light, making it appear more yellow, as I observed in Turkey in 1999 (Sun-bleached straw) and Egypt in 2006.

Eclipse earthshine

The corona's intensity can also mask another subtle eclipse phenomenon: earthshine on the Moon's night side. The brilliant light of the corona visually enhances the darkness of the new Moon's face, making observations of any lunar features invisible at a glance. Seeing such features requires a dedicated effort, and the sacrifice of time – a sacrifice many eclipse goers just don't feel is warranted, especially since earthshine is common during the Moon's waxing and waning phases every month.

But if you happen to be an eclipse veteran looking for a challenge, seeing earthshine is it! While a long enough exposure will clearly show eclipse earthshine, trying to capture it visually through binoculars will try your patience and test your observing skills. The Swedish astronomer Bigerus Vassenius made the first known observation of the phenomenon, informing the Royal Society that during the May 13, 1733, total solar eclipse, he detected several of the Moon's features through a telescope of 21-feet focal length – the optics of which were most likely inferior to the binoculars you use today.

Nineteenth-century observers made many "colorful" telescopic descriptions of earthshine during totality.

British observer Richard A. Proctor (1837–1888) of Cambridge published one of the richest accounts in a 1871 *Monthly Notices of the Royal Astronomical Society*:

> During the late eclipse the Moon's disk appeared green. It was compared by one observer to dark green velvet . . .
>
> It seems to me that a sufficient explanation is to be found in the nature of the light received by the Moon from the Earth during the Eclipse . . . Its colour must have been green, I think; because the proportion of land and sea surface in the terrestrial disk, as seen from the Moon, was such that calling the ocean blue-green and the land brownish (on the average), the resulting mixed colour would be a deep green . . .
>
> Now during the the Eclipse of 1860 land and sea were turned towards the Moon in different proportion . . . the two Americas were well advanced upon the Earth's disk as seen from the Moon. One would, therefore, expect that the Moon's disk on that occasion should have presented a brown hue. And accordingly we find Mr. De La Rue so describing it.

My observations have never revealed such colors. Then again, I am not one to judge, since I have yet to study the Moon's dusky face at totality through a telescope. But at the 1999 eclipse over Turkey, I did see lunar markings with the unaided eye through a Pringles™ potato chip can . . . I'll explain.

One of my main objectives at this eclipse was to see if any lunar features could be distinguished at totality with the unaided eye. Knowing that the feat would almost certainly be impossible owing to contrast effects, I devised a way to occult the corona. About a week before totality, when I was enjoying the historical sites of Turkey, I took a Pringles potato chip can out of my hotel's mini-bar, emptied the contents, cleaned the interior, then used a pocket knife to punch a tiny circular hole in the can's metal bottom. On the first clear dawn, I looked at the Moon through the pinhole, then used my knife again to adjust the hole until it neatly fit the circumference of the Moon's disk as seen through the open end of the tube. On eclipse morning, I purchased some black nylon stockings and created a hood sheer enough for me to see the corona sufficiently to point the tube but dark enough to block any scattered light from my surroundings.

The experiment was a success. With the corona blocked from view, I could see the Moon's face looking like a purple plum with black bruises. The bruised sections were the Moon's maria, while the purple regions were the lunar highlands reflecting earthlight. I have since used 10×50 binoculars to see lunar features at two more eclipses. In both cases, the Moon was purple and black. But the "bright" features wavered in and out of view as the eye roamed the disk in search of these elusive patches.

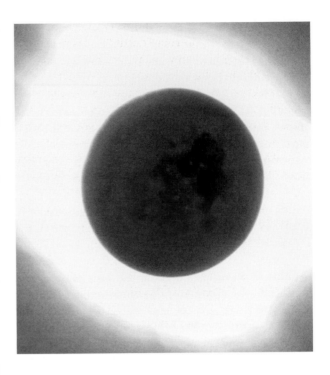

"Picturesque" observations

Before the sand of totality runs out, and the Sun's radiant limb breaks through the lunar valleys lining the Moon's western edge, be sure to allow a few seconds to appreciate the world around you as it is bathed in the Moon's shadow – and especially the colors of the landscape (everything shines with a platinum pallor) and the behavior of animals, plants, and other living things. During totality, it is quite common for birds to roost, and cows and other animals to become confused. During the 1851 totality at Gottenburg, "all sounds of labour ceased during the darkness." And Mr. Clarke noticed that the petals of a plant called the "'four o'clock flower,' from the time of the usual opening of its blossoms," began to bloom. "Three petals began to unfold," he noted, "but during the greatest obscuration they closed again; nor did they fairly unfold till the eclipse was nearly over."

My wife, Donna, made a wonderful "picturesque observation" during the 1991 eclipse from Hawaii, an account of which was published in the January 1992 issue of *Science Probe!* magazine. About 10 minutes after first contact, Donna grabbed her snorkel and fins and dove into the tepid waters of Waiulua Bay. There she observed the conditions of the undersea world before second contact. The most apparent observation was that, prior to eclipse day, she had snorkeled these waters and had seen "many species of fish . . . [that] come to eat, nest, and find shelter among the coral." But during the deep partial phases of the eclipse, "there were no fish. Nothing! . . . The only movement was a red and white juvenile yellow-tail coris hiding among some coral branches. All the fish were either buried, hidden in rocks and crevices or wherever they go when night approaches and biters come out."

Donna emerged from under the sea about five minutes before totality, but she was greeted by a tense atmosphere as clouds descended upon the Sun. After about three minutes into totality, Donna realized that "the water was the most interesting place to be so in I went. The coral had begun to bloom before I left and now were open, giving each branch a soft, feathered appearance. The exposed animals were delicate lavender and rosy pink. A few fish appeared as the sunshine returned. The fish were faster than the coral in response to the brightening, and some were able to nibble excitedly at the fleshy, exposed polyps before they pulled back into their protective limestone cups. Within 15 minutes things were back to normal." We can all learn a valuable lesson from Donna's observation: totality happens, clouds or not, so enjoy the *experience* – clouds or not.

The pre- and post-corona

I first bathed in the Moon's umbral shadow on October 2, 1959, roughly one month before my third birthday. Since then, I have seen the Sun's corona a dozen times – but not always during totality.

On March 7, 1970, I experienced an eclipse just outside the path of totality. From my home in Cambridge, Massachusetts, the Moon covered only about 97 percent of the Sun's diameter. Still, I could see the Moon's entire silhouette when I used my upheld hand to block the Sun's blazing crescent. What I saw was awesome: a pale lunar silhouette rimmed with pearly light. As I have described, it's common for eclipse watchers to see the Sun's inner corona up to a minute before second contact. That's because the corona is more than 100 times brighter near the Sun's limb than it is only one solar radius out. But during the last century and a half, observers have made notable observations of the Sun's corona in the minutes before and after totality – especially if the sunlit crescent is *entirely* blocked by a foreground object when viewed through unfiltered binoculars. **But you must be careful to keep the sunlit crescent out of sight at all times!** Careful observers have done this without any harm to the eyes.

During the eclipse of December 12, 1871, English solar astronomer Joseph Norman Lockyer expounded on the revelation of the early corona over Bekul, India:

> . . . the eclipse terminated for the others but not for me. For nearly three minutes did the coronal structure impress itself on my retina, until at last it faded away in the rapidly increasing sunlight.

Others had seen similar effects up to 1905, when, to my knowledge, S. Maitland Baird Gemmill, arguably made the last visual observation of this early corona until our generation. Professional interest in the visual appearance of the Sun's corona had all but vanished owing to the promise of the photographic plate. So powerful was the promise of photography that during that 1889 eclipse, Edward Emerson Barnard, the keenest visual observer of his day, "avoided being thrown off his [photographic]

program by simply not looking up at the Sun," writes William Sheehan in *The Immortal Fire Within: The Life and Work of Edward Emerson Barnard* (Cambridge University Press; Cambridge). This act, Sheehan notes, "left no doubt that the camera had supplanted the sketchbook in the role of historiographer of eclipses."

My interest in the phenomenon was kindled during the 1995 totality, when, after the occurrence of third contact, I noticed that the Moon's leading limb remained encircled by a faint ring of pale blue coronal light for more than a minute. But it was not until 2001 that I dedicated myself to seeing how long I could follow the corona after totality.[10]

I did this by strategically positioning myself so that the growing solar crescent was blocked by a tree branch. Dennis di Cicco did the same but with a video camera. Minutes ticked by with the corona undeniably visible. The silhouette of the Moon against the corona did not look black but charcoal gray. A thin ring of bright corona surrounded the Moon's silhouette; beyond it the corona appeared exceptionally feathery, extending some 10 arcminutes from the Moon's limb. I saw, and Dennis videotaped, the corona for 6 minutes and 54 seconds!

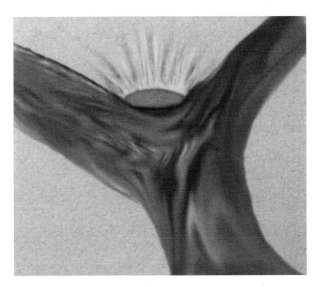

But this was far from being a record. During the July 18, 1860 total solar eclipse, a dozen observers reported seeing the corona either before or after totality, including Frederick Petit, director of the observatory at Toulouse, who saw it extending more than 3 arcminutes from the Sun's limb beginning 12 minutes before the start of totality. Likewise, during the July 8, 1842, total solar eclipse, Francois Arago, the French dean of astronomy, first caught sight of the corona 12 minutes before totality when the Sun was only 93 percent eclipsed.

Andrew Greenwood came close to breaking the pre-totality corona record during the 2006 eclipse; he spied it telescopically about 10 minutes before totality. "With the Sun's partial phase out of the field of view," he says,

---

[10] I focused my attention on this phenomenon after totality because I wanted to see other phenomena before totality, leaving me with ample time to search without other eclipse distractions.

"the corona was clearly visible. I now wonder how much sooner I could have seen it. It wouldn't surprise me if the corona could have been detected 20 minutes before totality under excellent clarity."

His estimate may not be far from the mark. From the Libyan Desert, near Salloum, Egypt, with the Sun at an altitude of 68°, I followed the corona with the unaided eye for 14 minutes after totality by placing the waxing crescent behind the intersection of two wooden tent poles. The corona was quite comfortably seen until only about a minute before the densest part of the corona turned into a whisker-thin frown that gradually faded away like Lewis Carroll's whimsical Cheshire Cat.

Annular eclipses

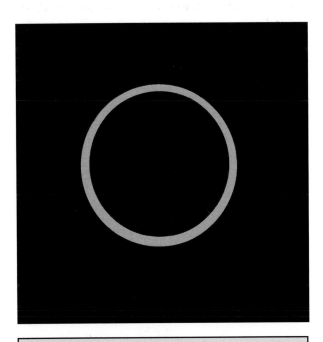

As I mentioned earlier, annular eclipses occur at or near lunar apogee, when the Moon is farthest from Earth; at these points in the Moon's orbit, we see the Moon smaller than the Sun in the sky, so it can't completely cover the Sun's face during a direct solar passage. Instead, we see the Moon's silhouette completely surrounded by a fierce "ring of fire." Although it may seem impossible, many eclipse phenomena can occur during annular eclipses. In fact, many eclipse phenomena were first discovered during annular events . . . including shadow bands!

The German astronomer and painter Hermann Meyer Solomon Goldschmidt appears to be the first person to document shadow bands, which he had observed from Frankfurt-on-the-Maine at the September 7, 1820, annular eclipse:

I was in my uncle's house, one hundred and twenty paces only from my father's, when the darkness began. I left the house; but I had scarcely gone twenty paces in the street, from East to West exactly, when I saw [the] moving shadows coming toward[s] me, covering the large and deserted street, and passing under my feet. Their movement was slow, for otherwise I could not have perceived them so distinctly, with the eyes so nearly fixed on them; the movement was not rapid. The aspect was like the shadow of smoke in sunshine – regular in translation; the forms, like rhomboids of four or six inches in diameter, and sometimes more extensive, were mixed with [superposed] shadows of the same size, and ribbon-like forms. The inner space[s] particularly, filled with round spots softly melted and mixed with the whole in veiled grey transparence, gave to the phenomenon a somewhat mysterious appearance. Thus, looking to the paved street, these ribbons and spots flitted under my feet, till I arrived at a group of men near my father's house, after having made one hundred steps in contemplating this strange apparition. At this moment the annular eclipse was forming: the light of the Sun ran around the Moon in a very singular manner – it seemed like a fluid mass. I can truly estimate the time from the moment when I saw the shadows till the forming of the annulus – an interval of two minutes and thirty seconds.

The diamond necklace, the chromosphere, and prominences, have all been seen at annular solar eclipses near the times of second and third contact. Indeed Francis Baily's famous account of "his" beads in 1836 details their appearance and behavior at an annular event. He and several other observers were captivated by the sudden onset of the luminous beads and their intervening black spaces during that eclipse, especially since they seemed to affect the timings of second and third contacts, which happened "between the moments when the beads of light were observed to appear and disappear, and when the black spots disappeared and appeared."

During the ring eclipse on May 10, 1994, the Moon covered only 94 percent of the Sun's face, but several observers saw and photographed all three phenomena. I saw this eclipse, in part, through 10 × 50 binoculars with solar filters, which made the Sun's photosphere appear a rich yellow. As the Moon's following limb neared the moment of second contact, the tips of the Sun's growing cusps turned blood red. Just then, a multitude of tiny black fingers poked into the Sun's western limb, shattering it into a tiny arc of yellow and red fragments. Then, in the instant before second contact, a single crimson prominence appeared all too briefly in the last black notch of the Moon before the Sun's limb reappeared.

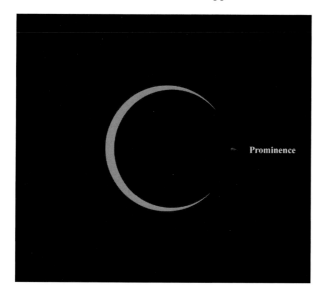

Seeing the prominence took me by surprise, but it shouldn't have. As Mark Littmann, Ken Willcox, and Fred Espenak note in their book, *Totality: Eclipses of the Sun* (University of Hawaii Press; Honolulu), "The first sure report of solar prominences came from Julius Firmicus Maternus in Sicily, who noticed them during the annular eclipse of July 17, 334."

More dramatic was the May 30, 1984, annular eclipse, in which the Sun and Moon appeared nearly the same size. From Greer, South Carolina, Leif J. Robinson and other members of the *Sky & Telescope* staff got to see 8 seconds of annularity. But Robinson may also have made history when he spied the Sun's corona at the trailing edge of the Moon's limb 73 seconds before central eclipse. Robinson kept the corona and a bright solar prominence in view for $2\frac{1}{2}$ minutes. Many of his companions saw, more or less, the same thing.

To my knowledge, no one had seen or photographed the corona during an annular eclipse prior to this event. But there have been subsequent sightings. During the February 1999 annular eclipse, which was visible for only 10 seconds from Australia, Fred Espenak photographed the Moon clearly backlit by the faint glow of coronal light. And during the June 2002 ring eclipse (99.6 percent), German amateur Friedhelm Dorst could trace the lunar limb against the Sun's inner corona for more than 7 minutes after maximum eclipse. Seeing the corona during an annular or hybrid eclipse requires extreme caution, since solar filters cannot be used for the observation; as described above for observations of the pre- and post-totality observations of the solar corona, observers need to block the uneclipsed portion of the Sun with a distant foreground object.

The emerald tiara

When a partial eclipse of the Sun occurs shortly after sunrise or shortly before sunset, all manner of curious phenomena may be seen on the Sun and Moon, owing to the density of the atmosphere along the line of sight, which can refract and disperse light. During these low-altitude events, the Sun and Moon can morph into misshapen forms and be fringed with curious spectral colors.

When the Sun is near the horizon, sunspots can be fringed with colorful light: red on the top and green on the bottom; again, the phenomenon is not a strange reversal in the way light is dispersed in our atmosphere but a perceptual novelty that catches the brain off guard and triggers a reversal in thought. This playful illusion is magnified when the Sun rises or sets partially eclipsed by the Moon.

Laurance R. Doyle [the Search for Extra Terrestrial Intelligence (SETI) Institute, Mountain View, California] brought this fact to my attention after observing from South Australia the sunset eclipse of December 4, 2002. As the partially eclipsed Sun neared setting, Doyle saw the Moon's lower limb with a distinct emerald green perimeter. He called it the "emerald-tiara" effect, which is in keeping with other jewelry-themed eclipse phenomena, such as the "diamond-ring" and "diamond- (or ruby-) necklace" effects.

As with the reverse red and green fringes observed in sunspots on the setting or rising Sun, when the Moon's lower limb partially covers the Sun at sunrise or sunset, the green edge at the bottom of the Moon's silhouette is actually every point along the Sun's "artificial" (concave) upper limb experiencing the normal effects of atmospheric dispersion near the horizon. Similarly, when the Moon's upper limb partially covers the Sun at sunrise or sunset, we may see the Sun's "artificial" (convex) lower limb naturally collared with red light. Remember, the cause of colorful atmospheric effects lies in how light (not darkness) interacts with the open air.

I used 10×50 binoculars to observe a stunning emerald-tiara effect during the October 13, 2004, sunset eclipse from Hawaii. As the advancing Moon nibbled away at the setting the Sun, the upper and lower rims of both the Sun and Moon displayed simultaneous complementary colors. As a bonus, an inky sunspot flashed "reverse" pink and green borders as the Sun and Moon exhibited their complementary green and red brims. These effects were magnified by intense mirages as the celestial pair set between the silhouettes of two billowing cumulus clouds.

Lunar eclipses

Unlike solar eclipses, no special filters are needed to protect the eyes during any phase of a lunar eclipse. They also last much longer, up to nearly three hours. Lunar eclipses are also visually much simpler events, and therefore more relaxing to watch; in other words, you don't have so many things to look for at once during the event, so you don't have to sacrifice seeing one aspect of the eclipse over another. And anyone on Earth who can see the Moon above the horizon at the time of the event can see the total phase eclipse; it is not limited to a narrow path on Earth, so you don't have to travel around the globe to see one – unless one is occurring beyond your horizons and you feel the need to see it.

But that hardly ever happens. Compared to the glory of a total solar eclipse, a total lunar eclipse is a more somber event – a slow death, if you will, of a romantic icon: the full Moon. We really don't get to see any new features on or off the Moon (like prominences and the corona during solar eclipses). We don't get to see daytime disappear, though we do get to see the night sky get darker than it normally would during full Moon.

Still, a lunar eclipse is an exciting and enjoyable event, especially when the eclipse is total. Watching the skull-white Moon being consumed by Earth's shadow, until it bleeds red at totality, can be an eerie experience. That certainly was the case in 1504 when provincial Indians in Jamaica encountered Christopher Columbus and his shipmates in ill health. The Indians, wanting nothing more than the men to starve to death or leave their homeland, had refused to offer food to the ailing Spaniards (in retribution for the cruelties they had suffered under the Spanish invasion). But Columbus devised a clever scheme to win over their affections. Knowing of a pending total lunar eclipse on March 1, the brave seaman gathered the local Indian chiefs together on the night before the event and told them, through a native interpreter, that their actions against the Spaniards had angered God, who was now determined to punish them with famine and pestilence. As a warning, Columbus said, God was going to make the Moon come out angry and inflamed.

The ruse worked. The next evening, the Moon slipped into the Earth's shadow, and fear and trepidation mounted among the Indians, who suddenly called for help, begging him to intercede. Columbus listened to the cries, then shut himself up while the eclipse was increasing. He did not reappear until totality had nearly ended. When Columbus came out of his room, he told the Indians that he had prayed for them and promised God on their behalf that they would be good and treat the Christians well. God, he said, had understood and forgiven the Indians . . . and promised to return the Moon to its former glory. Immediately after Columbus finished these words, totality ended, and the Indians saw the anger and inflammation pass away from the Moon. Certainly, this story inspired Mark Twain to write a similar episode in his wonderful novel, *A Connecticut Yankee in King Arthur's Court.*

While all solar eclipses occur at new Moon, all lunar eclipses occur at full Moon. As with solar eclipses, lunar eclipses do not occur every month because, as I previously explained, the Moon's orbit is tilted about 5° with respect to Earth's orbit around the Sun. Most of the time, the full Moon sails a little above or below the Earth's shadow. But up to three times a year – when full Moon occurs at or near one of the orbital nodes – the Moon can slip into Earth's shadow (either the dark, central umbra or less dark, outer penumbra) and be eclipsed by it.

Total lunar eclipses occur when the Moon fully enters the umbra. The eclipse is *partial* when the Moon partly passes through the umbral shadow. If the Moon misses the umbra but sails through the penumbra, we see a *penumbral* lunar eclipse.

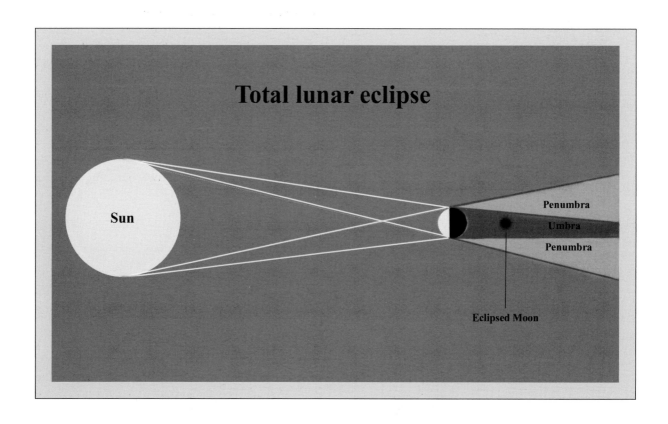

Total lunar eclipse

Sun

Penumbra

Umbra

Penumbra

Eclipsed Moon

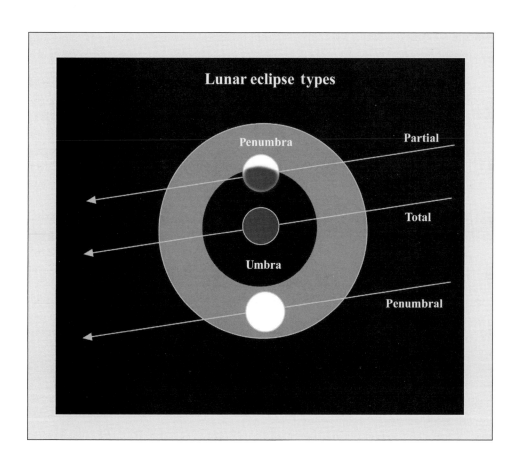

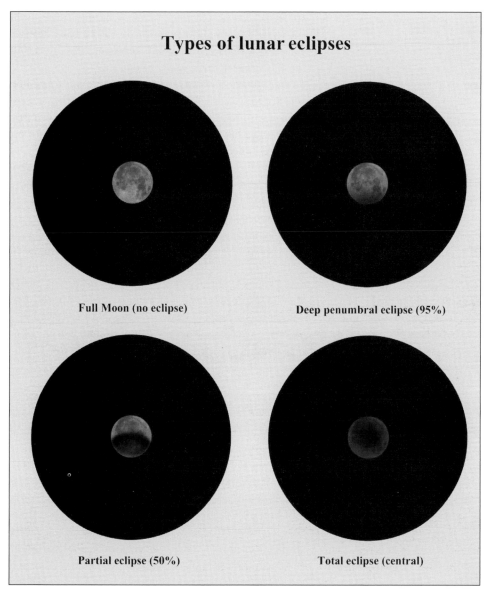

Penumbral eclipses

Penumbral eclipses are almost impossible to detect unless at least 60 percent of the Moon is eclipsed by the penumbral shadow. Even then, it takes an extremely astute observer with a trained eye to perceive the subtle change of light on the Moon's face at this point, which is something akin to it being eclipsed by a thin veil of smoke, causing a nearly imperceptible dulling of the Moon in the eclipsed regions. Deep penumbral eclipses (those greater than 80 percent) are easier to notice, especially through binoculars. Actually, trying to pick out the faint penumbral shadow against the full Moon's brilliant disk helps train the eye in its ability to detect subtle differences in light – a skill the best amateur astronomers strive to champion.

Extremely deep penumbral eclipses (those greater than 90 percent) are usually not a challenge to see even with the unaided eyes. The conditions of one's eyes, the state of the atmosphere, and the altitude of the Moon during greatest penumbral eclipse can all affect what you see or don't see.

My wife Donna and I were married on October 7, 1987 – the day of an extremely deep penumbral lunar eclipse (99 percent), which we easily saw without optical aid over the simmering caldera of Kilauea volcano on Hawaii's Big Island. The eclipsed Moon appeared to have an extremely dark nibble taken out of its southern limb; this nibble was also surrounded by a collar of dark gray penumbral fuzz. The "bite" was so noticeable that its form appeared on a shot taken with a Polaroid instant camera.

Partial eclipses

Partial lunar eclipses are exciting to watch, even without optical aid. The drama begins at first contact, when the umbral shadow takes a bite out of the Moon's eastern edge. As time passes, we see the curved and cheerless shadow creeping across more and more of the Moon's luminous face.

The umbral shadow appears darkest around the time of first contact. This is largely a contrast effect. But as the eclipse progresses, you should see, especially through your binoculars, that the shadow is not as black as ink. You can see the Moon shining through it. The shadow also appears colored. How can that be?

The solid globe of the Earth completely blocks the Sun's light. But some of the sunlight falling on Earth's atmosphere is bent around by it into the shadow. According to Richard Keen, a meteorologist at the University of Colorado, at the distance of the Moon, most of the light illuminating the Moon during an eclipse passes through the stratosphere, which lies 10 to 30 miles above the ground, and is reddened by scattering. That's why many lunar eclipses appear orange or red; it's the same light-scattering effect that causes our Sun to appear deep orange or red at the times of sunrise or sunset.

As Earth's umbral shadow covers more and more of the Moon, keen-eyed binocular observers should pay close attention to the shadow's leading edge. Soon after first contact, it may be possible to see a distinct blue or green collar hugging the dark shadow (which appears to become more and more transparent and redder) as time passes, and more and more of the Moon is eclipsed. Keen notes that this colored collar results from light passing through the upper stratosphere and penetrating the ozone layer, which absorbs red light – thus making that section of the shadow appear bluer.

But use your binoculars to watch for subtle color changes as the partial phases progress. The colors we see depend on a variety of factors, including how deeply the Moon enters the umbra, local atmospheric and contrast effects, and whether volcanic aerosols or other contaminants are in the stratosphere.

During the July 5, 2001, partial lunar eclipse, the northern half of the Moon was eclipsed by Earth's umbral shadow. Through 10 × 50 binoculars, I saw the Moon's northern highlands, which were deep in shadow, glowing with a burnt sierra sheen. The dark maria south of the highlands were copper, while the leading edge of the umbra shone with a pale gun-barrel-blue light; beyond it, the deepest regions of the penumbra appeared as a wash of ocher light.

The colors differed significantly the previous year when, from David Levy's house in Vail, Arizona, I saw the partial phases of a central total lunar eclipse on July 16, 2000. When the Moon was half eclipsed, the deepest portions of the umbral shadow appeared rusty gray. The umbra's outer edge was a somber gray band that gave way to a deep smog-yellow penumbra, which faded into the orange light of the low full Moon before it set into the dusty and distant landscape.

And shortly after sunset on February 20, 2008, Donna and I watched a half-eclipsed Moon leaving the southern portion of the umbra. This was all the more dramatic because the eclipsed Moon appeared inside the Earth's shadow projected onto the atmosphere. The section of the Moon in deepest shadow burned red. The shadow's edge was deep blue. And the penumbra had a deep ocher tint. All this was seen against the deep blue of the Earth's shadow projected onto the atmosphere, which in turn was capped by the rose-colored Girdle of Venus.

Optical illusions

It's true that, at least in part, the colors during an eclipse can be subjective, especially if simultaneous contrast effects are at play. But three other, and equally dramatic, visual illusions can occur during the partial phases. The first is *irradiation* – the apparent enlargement of bright surfaces adjacent to darker surfaces. We see this effect monthly whenever a young or old crescent Moon shines in the sky and we see the bright crescent appearing to bulge beyond the apparent limb of the Moon illuminated by earthshine. The ratio of the respective diameters is something akin to 6:5. Likewise, during the very early and late stages of the partial phases of a lunar eclipse, the bright uneclipsed portion of the Moon seems to swell beyond the limb of the eclipsed portion of the Moon.

The irradiation effect is a complex physiological one. In his book *Visual Illusions* (Dover; New York, 1965), Matthew Luckiesh tells us that part of it may be caused by a "rapid spreading of the excitation over the retina extending quite beyond the border of the more intensely stimulated region." He also notes that eye movements may also be responsible, as well as spherical aberration in the eye lens and the diffraction of light at the pupil.

Another illusion is the appearance of a jagged edge to the leading or following edge to the Earth's umbral shadow (see the photo below). This is quite apparent, especially to the unaided eye. It is not caused by irregularities in the Earth's shadow but by irregularities in the Moon's face. As the Earth's dark umbral shadow sweeps across the Moon, it blends with dark maria, which can appear to extend the shadow beyond its normal limits. The opposite effect happens when the Moon passes through the north or south parts of the umbral shadow; that's when its bright northern or southern highlands can appear to extend the cusps of the bright uneclipsed Moon, as seen with the unaided eyes. Most of these illusions vanish when seen through binoculars, which magnify and "diffuse" the shadows, decreasing contrast effects, and making the scene much more appreciable to the scientific eye.

Total eclipses

While the Moon's face can blush with color during a deep partial eclipse, the colors can pale in comparison to those seen during a total lunar eclipse. After second contact (the moment of totality), the Moon undergoes a slow series of transitions in intensity and color that is most magnificent when seen through binoculars. More than a telescope, the lunar eclipse belongs to the realm of binocular observing – at least in the arena of aesthetic appeal.

But, like total solar eclipses, total lunar eclipses have also had an affect on humanity throughout the ages and sometimes on history. During the Peloponnesian War, for instance, Athenians saw the total lunar eclipse of 413 BC as a bad omen and delayed their planned retreat; this ultimately led to the death of the Athenian general Nicias and the ruin of his army in Sicily . . . which ultimately led to the fall of Athens, the once shining jewel of Greek civilization.

Lunar eclipses were long seen as a sinister sign – one of a divine disturbance in response to mortal displeasure. In India, the Brahmins of old thought that when the Moon was eclipsed, she was fighting with a black devil. Aristotle believed that lunar eclipses were a possible cause of earthquakes. In his diary, Samuel Sewall, alludes to a connection between the birth of a deformed child ("one whose tongue had grown fast to the roof of the mouth," among other deformities) and a portentous event: "a bloody-coulour'd Eclipse of the Moon, onely middle of the upper part of a duskish dark."

Many ancients also saw the progression of Earth's shadow across the face of the Moon as an act of devouring, which, during the climax of totality, magnified the gory drama, as the virgin purity of the Moon was stained with blood. The blood red aspect of totality also gave rise to the belief that the Moon could bleed like a woman.

Color shifts

The more eclipses you see, the quicker you'll discover that not all total eclipses are blood red; in fact, these seem to be in the minority, which may explain their ominous hold on the human imagination. Sometimes the totally eclipsed Moon is more yellow or orange. More often than not, it appears to burn like hot copper. These colors can also be complemented by a wide variety of shades – blue, green, aqua, peach, and cream – any of which can come and go with the Moon's passage through the umbral shadow.

As I mentioned earlier, the color one sees depends on a number of atmospheric and optical effects. The color may also change, sometimes dramatically, as the eclipse progresses. Color shifts are most remarkable during totality. And few total lunar eclipses ever look quite the same. As a rule of thumb, the more transparent Earth's upper atmosphere, the brighter and more colorful the eclipse. The photo on page 66 shows the bright total lunar eclipse of August 27–28, 2007, around mid-totality. During this eclipse (a beautiful orange one), the Moon's northern limb skirted the central portion of the Earth's umbral shadow. Note the brightness difference between the Moon's northern and southern limbs.

Sometimes, the Moon is so dark that it all but vanishes from the sky. This was clearly the case during the October 4, 1884, total lunar eclipse, which occurred a little more than a year after the great Krakatau eruption in August, 1883. In the Royal Society's 1888 publication, *The Eruption of Krakatoa* (Harrison and Sons; London), Edmond J. Spitta of Clapham, London, is quoted as saying that "During totality the [M]oon was, generally speaking, exceedingly faint – indeed, at times barely visible to the naked eye – and presented none of the coppery colour usual on these occasions." And the Rev. S. J. Johnson of Bridport reports that at the middle of totality, "To the naked eye nothing could be seen but a faint nebulous spot . . . The exact appearance of the Moon at this time would be described by quoting Kepler's words verbatim about the eclipse of June (not [D]ecember) 1620. 'Luna difficillime apparuit, emicuit tamen instar tenuissimae nebeculae, longe debilior quam via lactea, sine omni rubedine.'"[11] Kepler's eclipse occurred in the year after the great July, 1619, eruption of Hekla in Iceland.

And during the March 30, 1885, total lunar eclipse, A. B. Briggs of Tasmania says (as quoted in the *Journal of the Royal Astronomical Society of Canada*, 1946, column 40, page 267) that, at the time of maximum, "All within the shadow was utterly obliterated – lost in the dead slaty tint of the sky. I could not distinguish a single crater after once it was fairly within the shadow. Not the slightest trace of the coppery tint was visible throughout."

When the Moon "disappears" during totality, it is usually the result of one of those rare occasions when mighty volcanic eruptions toss huge quantities of dust and sulfuric aerosols into the stratosphere. When sulfur dioxide gas is thrust into the stratosphere by a powerful volcanic blast, it first passes through Earth's cold lower atmosphere. As the gas rushes upward, it reacts with water droplets and sunlight to form aerosols of sulfuric acid, which condense rapidly in the stratosphere. While the ash generally settles out, the sulfuric-acid aerosols stay in the stratosphere for years. (The rule of thumb is that one-third of the aerosols settle out each year.) This sulfuric acid haze effectively reflects, scatters, and absorbs sunlight, severely dimming sunlight and causing dark eclipses.

After the powerful June eruptions of Mount Pinatubo in the Philippines in 1991 (which erupted millions of tons of sulfur dioxide into the stratosphere), the first total lunar eclipse to occur after it (on December 9, 1992) was so dark that the Moon's disk was barely visible over the Boston skyline, with only a small portion of its northern limb dimly visible to the unaided eyes. Had this been a central eclipse, the Moon probably would have vanished.

Danjon scale of brightness and color

How bright or colorful is a total lunar eclipse? Many skilled eclipse observers use a five-point scale created by French astronomer André Danjon (1890–1967) in the 1920s. Known as the Danjon scale, it is a simple and effective way for naked-eye and binocular observers to rate the appearance of an eclipse at mid totality. All you have to do is compare what you see with Danjon's luminosity (L) values and descriptions below:

Danjon scale of brightness and color

L value	Description
0	Very dark eclipse, Moon almost invisible, especially at mid totality.
1	Dark eclipse, gray or brownish in color; surface details on the Moon can be seen only with difficulty.
2	Deep red or rust-colored eclipse. A very dark patch appears in the central part of the umbral shadow. The outer umbra is relatively bright.
3	Brick-red eclipse. The umbral shadow usually has a bright or yellow rim.
4	Very bright copper-red or orange eclipse with a very bright, bluish umbral rim.

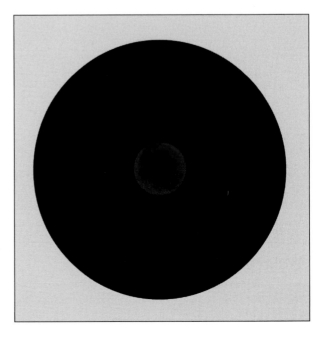

[11] . . . nothing could be seen of it, though the stars shone brightly all around.

If possible, make Danjon estimates at the onset of totality, at mid eclipse, and at the end of totality. The image at bottom right on page 66 shows a dark eclipse at mid totality. The brightness and color changes, as I said, can vary dramatically. During the August 16–17, 1989, total lunar eclipse, for instance, I used my unaided eyes and recorded these three Danjon values and descriptions (note too the fractional values):

	L value	Description
Second contact:	2.7	*Red-brown center with a dusky gray-blue patch. Maria dimly visible. Moon's eastern rim a mixture of straw, orange, copper, red, and pale purple. Western rim pale purple*
Mid totality:	2.3	*Dark red to chocolate brown umbral core that thins to rust-orange, then to yellow-orange and straw toward the Moon's western limb. Eastern limb: orange.*
Third contact:	3.2	*Small, dark gray-blue umbral core. Moon's eastern limb: blue. The Moon's western limb has a fantastic array of bright colors: yellow-white, straw, tangerine and purple-orange.*

You can also judge the Moon's brightness at mid-totality by making a reverse-binocular magnitude estimate. Just turn your binoculars around and look at the Moon through one of the objectives lenses with one eye. Now compare this reverse view with the naked-eye view of bright stars of known magnitude. You will have to adjust your estimate, though, by factoring in the reducing power of your binoculars; 10× binoculars will diminish the Moon's apparent brightness by a factor of 100, or 5 magnitudes.

So, if you see a very dark eclipse, say, 5th magnitude as seen through your reversed 10 × 50 binoculars, the actual magnitude of the Moon is equal to that of a 0-magnitude star. Binoculars that magnify seven times, reduce the Moon's light by a factor of 4.2 magnitudes. The photo illustration below shows how the Moon should appear as seen through one objective lens using the reverse-binocular method.

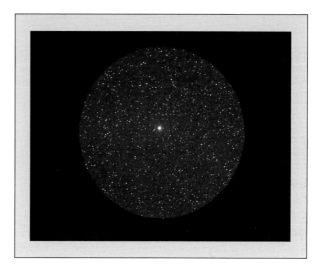

No matter how you decide to watch or monitor the spectacle, don't forget to relax and have fun!

CHAPTER 4

The planets: worlds of wonder

To know about the planets is to know about ourselves.
— Martha Evans Martin, *The Way of the Planets* (1912)

When you look up at the night sky, do you ever wonder about the stars? Our distant ancestors did. They revered these untouchable flames, believing that some spiritual force governed their solemn march across the sky. They worried when certain stars vanished beneath the western horizon (the portal of death) and rejoiced when they rose again in the east (the gateway to life). They noticed the clockwork regularity of the fixed stars and marked the seasons by their passages. Eventually, ways were found to make the stars more familiar. Names were given, patterns fashioned among them, and myths created to explain their placements in the heavens.

Early skywatchers also noticed that not all of the naked-eye stars were fixed. Five of them wandered. The Greeks called them *planets* (meaning "wanderers") and named them after gods – Mercury, Venus, Mars, Jupiter, and Saturn. Like the Sun and Moon, the planets moved along an invisible pathway in the sky (the *ecliptic*), which coursed through the 12 constellations (or signs) of the zodiac. Two of them (Mercury and Venus) threaded the twilight skies close to the Sun, never making it into the deep-dark night. The other three planets (Mars, Jupiter, and Saturn) ventured deep into the night . . . but in mysterious ways: aside from marching stately forward (eastward), they would, at times, temporarily stop their eastward course and then reverse, before moving eastward again.

Ancient Babylonians imparted occult significance to these motions – as if the wanderers were writing messages in a secret language. Cracking the code became tantamount to communing with God. The planets were not "gods" themselves but manifestations of the gods – through which their will could become known. Clearly, they reasoned, since Earth was at the center of all heavenly motion, human beings must be the center of His attention.

This age-old wondering has not been lost on stargazers of today. Consider these words from planet observer and prolific author William Sheehan:

> There certainly is something deeply moving about the "pageant" of the heavens. Throughout my life, the ascent of Venus from behind the Sun, climbing into the sky, and then descending – rapidly – back toward the Sun, the planet has created a rhythm that has structured my own life. In some ways one sees in the phases of Venus's movement something along the lines of the stages of the life of man. Birth; the long slow development of childhood; one's ascent into the prime of life – when one stands proudly on the forehead of heaven and seems impregnable and as if one will stand there forever; then one's rapid descent and decline into death. I have been watching Venus this past [2008–9] apparition with as much interest as when I was a youth. There is something archetypal about the pageant that is quite independent of what we know about Venus – its motion around the Sun.

It took humankind nearly 5,000 years to break its egocentric view. Today we know that the Earth and Moon belong to a vast family of objects gravitationally bound to the Sun (*Sol*, to the Romans) – an unassuming, G-type, dwarf star marking the gravitational center of our *Solar System*.

Although you cannot feel it, our world is spinning at 1,000 miles per hour and moving through space at 1,000 miles per second. The Moon is whirling too, revolving around the Earth once a month, as the Earth orbits the Sun once a year. All the planets join us in this carnival ride through space, each moving at a speed determined by its distance from the Sun; the closer a planet is to the Sun the faster it revolves around it.

Most of us grew up knowing that our Solar System had nine planets, plus a bevy of planet-circling moons, thousands of small rocky bodies called asteroids (most in the Main Asteroid Belt between Mars and Jupiter), graceful comets, and countless particles of interplanetary dust. The nine planets were the main players in this cast, and here they are in order of increasing

distance[1] from the Sun in millions of miles: Mercury (36), Venus (67), Earth (93), Mars (142), Jupiter (483), Saturn (886), Uranus (1,783), Neptune (2,791), and Pluto (3,671).

We also divided this Solar System into two equal parts: the inner and outer Solar System. Mercury, Venus, Earth, Mars, and the Asteroid Belt comprised the inner solar system. Jupiter, Saturn, Uranus, Neptune, and Pluto ruled the outer system. We called the four largest inner planets (Mercury, Venus, Earth, and Mars) the *terrestrial* planets, after the Latin word *terra*, meaning Earth; they are made primarily of rock and metal. We knew the four largest outer planets (Jupiter, Saturn, Uranus, and Neptune) as the gas-giant, or Jovian, planets; they are composed primarily of hydrogen and helium with a metal or rocky core. But what about Pluto?

Ever since Clyde Tombaugh discovered Pluto in 1930, astronomers looked upon that world as a misfit. For instance, all the planets, except for Pluto, orbit the Sun in nearly circular paths, in a plane formed by the imaginary extension of the Sun's equator into space. Pluto's orbit, on the other hand, is not only steeply inclined 17° to that plane but also highly elongated (cigar shaped).

Because of its odd-shaped orbit, Pluto actually spends about 20 years inside the orbit of Neptune; no other planet crosses the orbit of another planet. Astronomers struggled to make Pluto fit into current Solar System models. But no one knew what this tiny ice world (about the size of Mercury) was doing beyond the gas-giant planets. Perhaps Pluto was not an outsider but one of many unseen trans-Neptunian planets. With this idea in mind, some astronomers began dividing the Solar System into three sections: the inner rocky planets, the middle gas-giant planets, and an outer realm of icy, trans-Neptunian worlds.

The latter scheme was prescient. Thanks to spacecraft exploration and advances in telescope technology, we rapidly expanded our knowledge of the deep Solar System. We learned that Pluto is only two-thirds the size of our Moon, which is less than two-thirds the size of Mercury. We guessed wrong because Pluto is not a single object but a world with three moons: Charon, Hydra, and Nix. We have also added to the Solar System thousands of new icy objects orbiting the Sun beyond the orbit of Neptune, in a zone called the Kuiper Belt. At least 100 of these new objects share similar, highly elliptical orbits with Pluto. Most exciting, however, was the discovery in 2003 of two new worlds (Sedna and Eris), which are three times more distant than Pluto: Sedna is slightly smaller than Pluto, while Eris is slightly larger than it. Pluto, then, no longer dominates this distant ice

[1] The distances are average distances, because the planets do not round the Sun in circular orbits but in oval-shaped curves, called ellipses, so each planet's distance from the Sun changes slightly as it orbits. The Earth, for instance, is closest to the Sun in July (91.5 million miles) and farthest away in January (94.5 million miles).

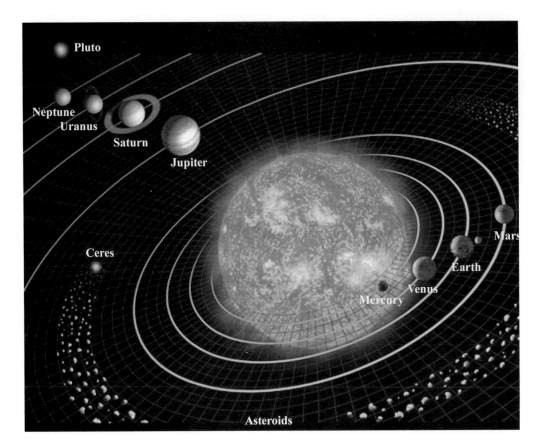

kingdom. Equally important, a surprising Hubble Space Telescope observation in 2005 revealed that Ceres, the largest known asteroid, is not only nearly round like a planet, but it also shares characteristics of the rocky terrestrial planets.

The new Solar System

We are living in exciting times. Based on the findings mentioned above and other data, the International Astronomical Union (IAU) adopted in August 2006 new definitions for planets and other Solar System bodies, placing them into three distinct categories: (1) *planets*, (2) *dwarf planets*, and (3) *small Solar System bodies*. According to the IAU's new edict:

(1) A *planet* is a celestial body that
    (a) orbits the Sun,
    (b) is nearly round, and
    (c) has cleared the neighborhood around its orbit.
(2) A *dwarf planet* is a celestial body that
    (a) orbits the Sun,
    (b) is nearly round,
    (c) has *not* cleared the neighborhood around its orbit, and
    (d) is not a Moon (a natural satellite orbiting a planet).
(3) *Small Solar System bodies*
    All other objects (except satellites) orbiting the Sun.

As of August 2006, our new Solar System consists of eight *planets* (Mercury, Venus, Earth, Mars, Jupiter, Saturn, Uranus, and Neptune), three *dwarf planets* (Eris, Pluto, and Ceres – though several more objects are being considered), and a multitude of *small Solar System bodies* (currently including most of the Solar System asteroids, most objects orbiting the Sun beyond the orbit of Neptune, comets, and other small bodies – like interplanetary dust). Since Tombaugh discovered Pluto, and in less time than it takes Uranus to orbit the Sun, a new Solar System has unfolded before our eyes.

But we should not become enamored with titles. To the ancients, the Sun and Moon were planets. So too were comets. And, as I discuss on page 99, in the nineteenth century, the number of planets in our Solar System expanded and contracted no less than five times. Perhaps it would be easier just to look at all these different objects as "Captives of the Sun," a phrase first coined by the late Thomas D. Nicholson, an astronomer at the American Museum of Natural History in New York. Whatever you feel, as new discoveries occur, and as our knowledge expands, the definitions of planets and other Solar System bodies could, and will in all likelihood, change again in the near future.

Sheehan agrees: "I find that all the debate about whether Pluto should be a planet – or not be one – is beside the point. Obviously these definitions are bound to be arbitrary to some extent. As Othello says – 'it's chaos come again.' Our definitions are messy because the Solar System is messy. It sort of grew like Topsy. It is the product of a lot of contingent accidents and haphazards."

What does all this mean for the binocular user? It means that, unless the tide of progress changes, *all* eight planets, one dwarf planet, and several small Solar System bodies,

Constellations of the zodiac

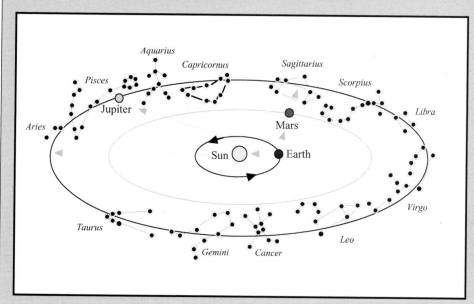

The zodiacal constellations are drawn as if seen from outside. Those in the foreground appear reversed. As viewed from Earth, Mars is in the zodiacal constellation of Sagittarius, Jupiter is in Pisces, and the Sun is on the cusp of Aries and Taurus.

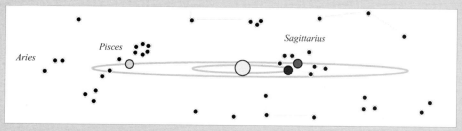

This more oblique view shows why we never see the Sun and planets against constellations that lie high above or below the plane of the Solar System. The section of sky where the plane of the Solar System intersects the celestial sphere is called the ecliptic plane or the ecliptic. The zodiacal constellations lie along this great circle.

are within reach of your simple instrument. So let's go out and explore these brave new worlds.

Worlds in motion

The planets orbit the Sun in a counterclockwise direction, along a plane that more or less parallels the imaginary one extending out from the Sun's equator; it is a consequence of orbital angular momentum in the solar nebula – the cloud of dust and gas from which the Sun and its family of worlds formed some 5 billion years ago.

Consequently, from our platform on Earth, we see the Sun, Moon, and planets sail across the heavens in a narrow corridor of sky called the *ecliptic*, or *zodiac*. This 360°-long band measures about 18° wide and frames the 12 zodiacal constellations: Aries the Ram, Taurus the Bull, Gemini the Twins, Cancer the Crab, Leo the Lion, Virgo the Virgin,

Libra the Scales, Scorpius the Scorpion, Sagittarius the Archer, Capricornus the Sea Goat, Aquarius the Water Bearer, and Pisces the Fish. Zodiac means "circle of animals"; the only the exception is Libra – a late addition to the circle, designed to even out the formerly 11 zodiacal constellations. As the planets move through these 12 "signs," they can alter the appearance of their familiar star patterns and confuse beginning stargazers.

You don't have to know how to see the zodiacal constellations to pick out the bright naked-eye planets, but it does help; it is imperative to know these star patterns to find the dimmer Solar System members with your binoculars. One way to learn these important constellations is to use my 2008 companion book *Observing the Night Sky With Binoculars* (Cambridge University Press; Cambridge). You could also purchase a star wheel, which shows the appearance of the stars and constellations at any hour on any day of the year; it will show the path of the ecliptic through the zodiacal constellations. It is along this line that the planets will appear.

Because the planets are constantly changing their orbital positions – and, therefore, their locations in the sky as seen from Earth – my telling you where to look to find them on any given date is beyond the scope of this book. Fortunately, finding the information is easy. Monthly astronomy magazines, local astronomy clubs and planetariums, and newspapers can provide you with that information. Some television programs – like Jack Horkheimer's Star Gazer – will tell you where and when to look for planets. Computer software, such as Starry Night® is invaluable: "all it takes is the click of a mouse and you'll learn all about the stars and planets shining over your house." The world wide web is also rich in sites with planet-finder information and free planetarium software (e.g. http://freeware.intrastar.net/planetarium.htm).

Planets usually give themselves away because they don't twinkle like the stars. They shine by reflected sunlight and are near enough to show tiny disks with optical aid. Stars are so distant that their light appears as point sources. More than disks, point sources are greatly subjected to turbulence in Earth's atmosphere, which continually refracts, or bends, the light into new directions. These light shifts can also magnify or dim the star's brightness, or cause it rapidly to change in color as different wavelengths are bent to varying degrees.

We see all this action as scintillation, or twinkling. (You can magnify twinkling effects by training your binoculars on a bright star low to the horizon; it's quite entertaining and fascinating to watch.) A planet's disk essentially averages out the twinkling effects, so we see it burning forth with a steadier glow; but you can see a planet perform the twinkling actions mentioned above (albeit more slowly than a star), especially when it is low in the sky. Indeed, one early Greek name for Mercury, which is always seen near the horizon, was Stilbon, meaning scintillating.

With practice and time, you can easily identify the naked-eye planets, because each moves at a distinct pace. The closer a planet is to the Sun, the faster it "rounds" the Sun. The table below compares the orbital periods of the major planets, rounded off to the nearest day or year:

Orbital periods of the major planets

Inferior planets	Orbital period
Mercury	88 days
Venus	225 days
[Earth	**365 days]**

Superior planets	Orbital period
Mars	2 years
Jupiter	12 years
Saturn	29 years
Uranus	84 years
Neptune	165 years

While Mercury may zip through a zodiacal constellation in a matter of days or weeks, the outer planets will take years. It's fun to plot (or image) a planet's position against the background stars and record its apparent motion.

Each planet also has defining visual characteristics. For instance, only the Sun and Moon outshine Venus. Mars, Jupiter, and Saturn can dominate the naked-eye midnight sky; Mercury and Venus cannot. Mars is our red planet. Jupiter has a golden hue. And Saturn shines with the light of warm straw.

The inferior planets

Mercury and Venus are inferior planets, meaning they orbit the Sun inside Earth's orbit. Consequently, we never see them stray very far from the Sun. They can only achieve a maximum angular distance, or greatest elongation, east or west of it.

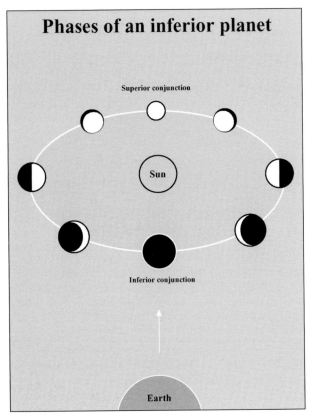

If you are a dedicated sky watcher, it's not difficult to spot an inferior planet. One day, it suddenly seems to appear in the twilit sky just above the western horizon after sunset, or just above the eastern horizon before sunrise. Gradually, day by day, it distances itself from the Sun – growing in prominence as it does – until it achieves greatest elongation, when it is most obvious. After that, the planet loops back toward the horizon from which it rose, then disappears below it. Weeks later, the planet suddenly reappears on the other side of the Sun, where it performs a similar dance.

So inferior planets seem to oscillate from one side of the Sun to the other. When inferior planets pass between the Earth and Sun, we say they are at inferior conjunction; when

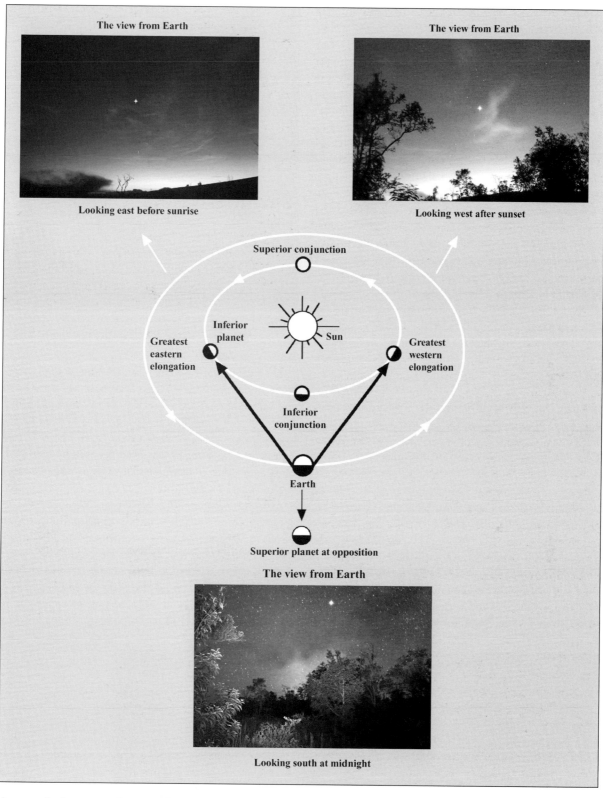

The view from Earth

Looking east before sunrise

The view from Earth

Looking west after sunset

Superior conjunction

Inferior planet

Sun

Greatest eastern elongation

Greatest western elongation

Inferior conjunction

Earth

Superior planet at opposition

The view from Earth

Looking south at midnight

they are farthest away from Earth, opposite the Sun, we say they are at *superior conjunction*. Only the *superior* planets, those that orbit the Sun outside Earth's orbit, completely circle the sky, passing successively through each of the zodiacal constellations (see the diagram on page 71). Only superior planets can be seen opposite the Sun as seen from Earth (*opposition*); at this time they will burn brightly above the southern horizon around midnight.

Mercury and Venus will, over time, display a series of phases like those of the Moon. Again, the phenomenon is due to their unique position inside Earth's orbit. What phase we see depends on the geometry of the situation (see the diagram above). When an inferior planet is closest to Earth, it lies between the Earth and the Sun, and its unilluminated side is facing us (as with new Moon). The crescent phases around inferior conjunction are most difficult to detect in binoculars because the illuminated crescent is razor-thin seen against bright twilight glow. In its gibbous phase, three-quarters of the planet is illuminated, but it is also farther away, making its disk smaller and its brightness more diminutive. The best time to see an inferior planet is when it is at one of its greatest elongations.

Of the two inferior planets, little Mercury presents planet watchers with the biggest challenge. Being the closet planet to the Sun, Mercury sets shortly after the Sun, or rises shortly before it. It is one of two true morning or evening stars (the other being Venus). You'll never see the planet in the dark sky, only in the twilight, the glow of which lowers the planet's contrast, making it more difficult to see than if it were to be seen against a dark sky, such as during a total solar eclipse.

Because of its low twilight position, the planet must also shine through the densest parts of Earth's atmosphere, which, through the process of light scattering by fine atmospheric particles, further dims its light – an effect called *extinction*. So let's start our planet discussions with this shy little devil... what Jack Horkheimer, director of the Miami Space Transit Planetarium in Florida, calls the pink-iron planet.

Mercury

Mercury is a tiny world, about the size of our Moon, with an aluminum silicate surface and a core twice as rich in iron than the Earth's. It also looks like the Moon. Broad plains, steep cliffs, and innumerable craters cover its surface. But an astronaut wouldn't want to stand there. Mercury is in the Solar System's hot seat. Its daytime temperatures rise to a scorching 450 °C (840 °F) – hot enough to melt tin or lead. Slip over to the dark side and the temperature plunges to −170 °C (−275 °F). The gravity of this little world (about one-third that of Earth's) is not strong enough to hold a substantial atmosphere. The few gases that do surround the planet (from the solar wind and passing comets) are so tenuous that they cannot scatter sunlight effectively enough to produce a tinted

sky – a Mercurian poet's nightmare, but a Mercurian astronomer's dream.

The Sun, appearing some three times larger than it does from Earth, simply rises crisp and sharp against a bleak and barren Mercurian landscape, only to slip unimpeded into a glorious black, star-studded sky. And while it's highly improbable that life exists on Mercury, its axis is so upright that water-ice has been detected by spacecraft in its deep polar craters, where it remains shielded from the Sun's searing rays.

A year on Mercury lasts only 88 Earth days. But it takes about 59 days (two-thirds of the orbital period) for the planet to turn once on its axis. A hypothetical Mercurian, then, would have to wait 88 Earth days for the Sun to set after rising, then wait the same period of time for it to reappear. In other words, Mercury does not keep the same face toward the Sun at all times, as was once thought by early telescopic observers.

Meet Mercury

Maximum brightness:	−1.9 (0.0, during average opposition)
Diameter:	4,879 km (3,032 miles)
Apparent diameter (arcseconds):	13.0 (max.); 4.5 (min.)
Mean relative density (water = 1):	5.4
Weight; if 100 lb on Earth:	38 lb
Mean distance (from Sun):	58,000,000 km (36,000,000 miles)
Mean orbital velocity:	48 km (30 miles) per second
Surface temperature (daytime):	450 °C (840 °F)
Period of revolution:	88 Earth days
Length of day:	59 Earth days
Orbital inclination:	7°
Axial tilt:	0°
Number of moons:	0
Rings:	No

Mercury is named for the swift-footed messenger of the Roman gods – a deity borrowed from the Greeks, who had first named their olympian messenger, Hermes. Because Mercury sprints around the Sun, the planet comes and goes from view quite quickly. To the unaided eye, it's visible for less than an hour after sunset, or before sunrise. The rapid oscillations between Mercury being the evening and morning star caused some confusion among early Greeks, who believed that these performances were the actions of two worlds, not one. The Egyptians, who called the world *Sobkou*, after their messenger god, were the first to discover that this "star" traveled around the Sun.

Most of the time, the planet hides in the glare of the Sun and is extremely difficult to detect, even with binoculars. The best viewing times are around its greatest eastern and western elongations, which come about six times a year. Owing to orbital circumstances, the most favorable elongations occur in the spring (March and April) and fall (September and October), when the planet appears "highest" in the sky from mid-northern latitudes. Of these, the most favorable occur when Mercury is at its greatest distance from the Sun in its orbit.

Mercury travels around the Sun in a highly eccentric orbit. When closest (at *perihelion*), Mercury is only 46 million kilometers (29 million miles) from the Sun; when farthest away (at *aphelion*), it is at 70 million kilometers (43.5 million miles). Consequently, at perihelion, we see Mercury only 18° from the Sun at maximum elongation; that angular distance changes to 28° when Mercury is at aphelion.

The great variation in Mercury's distance from the Sun also affects its brightness (as does the phase of the planet).

Mercury attains a maximum brightness of magnitude −1.9. And though that makes the planet a little bit brighter than Sirius, the brightest star in the sky, the planet is *very* near the Sun at superior conjunction at that time, making it extremely difficult (and dangerous) to observe. And while the planet is faintest (∼ magnitude 0) at greatest elongation, it is then seen against a darker twilit sky, making it much easier to see. Mercury at greatest elongation is about as noticeable as golden Arcturus is at dusk or dawn. (Also, you have to remember that you're seeing a planet about the size of the Moon, being illuminated by reflected sunlight some 57 million miles distant; the Moon is only 250,000 miles away!)

With binoculars you can follow Mercury long into the twilight, especially at western elongation when the planet emerges at dusk before the Sun. Simply line up the planet with some familiar terrestrial object nearby and wait for the Sun to rise. You'll find that as the sky gets brighter and brighter, Mercury becomes more and more difficult to pick out from the background sky. The planet's brightness will appear to dim, slowly at first, then rapidly; the planet's light is not really fading, it's just being visually lost as the contrast between it and the surrounding sky lessens; until it's like trying to spot a polar bear in the snow. (You'll also discover that without additional starlight or anything in the field, it's hard to focus your eyes on a seemingly blank sky.)

At its best elongations, Mercury can be followed into the early daylight through binoculars, but this requires constant vigilance. (I discuss the problem of seeing planets in the daytime in greater detail on page 81; I also give helpful hints on how best to achieve success.) Sometimes nature aids you in your search – namely, on occasion, the Moon and Venus (the two brightest daytime celestial objects other than the Sun) pass near Mercury, helping you to pinpoint the exact location of the planet and thereby to focus your attention on the right spot in the sky; you can then focus your eyes.

I know of only one person to have spied Mercury in the daytime with the unaided eye without the aid of a solar eclipse. On June 28, 2004, Becky Ramotowski of Tijeras, New Mexico, made her historic observation at at an elevation of 7,205 feet. At the time, Mercury had sailed close to brighter Venus before sundown. The drama began 53 minutes before sundown, when she found Venus and Mercury in her 8 × 56 binoculars. While she could see Venus with her unaided eyes, she had to wait until 12 minutes before sunset to catch a fleeting, though not certain, glimpse of Mercury. Then, at 8:17 (6 minutes before sunset), Ramotowski made history. After cupping her hands to her eyes, she saw Mercury as a definite "tiny sparkle" at the same spot she had suspected it before.

I described Ramotwoski's observation in more detail in the April 2004 issue of *Sky & Telescope*. I also quoted Bradley E. Schaefer, an astronomer at Louisiana State University, who put Ramotwoski's observation into historical perspective: "Ramotowski's sighting was close to the edge," he said. "She had full modern knowledge and the aid of binoculars, so we now know that sighting Mercury in the daytime can be done, but only under the best of conditions." Schaefer knows of no ancient records of Mercury being spotted in the daytime, but the absence of evidence does not mean evidence of absence. "People long ago were persistent and smart," he says, "so I readily think that Mercury has been occasionally spotted in daytime over the past many centuries."

It is extremely difficult to detect any phase but the crescent phase of Mercury with ordinary handheld binoculars, such as 7 × 50 or 10 × 50 binoculars. Large, high-quality binoculars, such as 25 × 100s mounted on a tripod, however, will show all the phases. Still, it takes time, patience, excellent focus, and the ability to discern a small effect on a tiny disk. With a casual glance, the planet will simply appear suspiciously "out of round." But it is possible.

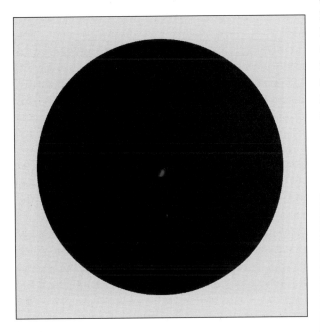

Perhaps the most pleasing attribute of this little world is its distinct pink or rose color. (Now you know why Horkheimer calls Mercury the pink-iron planet – it's the pink planet with a dense iron core.) It's a subtle shade, like the flush of embarrassment. Certainly, one could argue that the color one sees is simply an effect of dust and other atmospheric pollutants in the low line of sight – the same effect that makes a low Sun or Moon appear orange or red. But light extinction (also called reddening) does not explain the planet's rosy hue when seen high in the sky through large binoculars and telescopes during the daylight.

What's both awesome and amazing is seeing Mercury during a total eclipse of the Sun. Then one can appreciate the brilliance of this diminutive world against a higher-contrast sky; it also appears dazzlingly white, not pink – an effect of high-contrast light falling on the eye's color-insensitive rod cells, which we use to see as darkness descends.

It's been rumored that a city dweller cannot see Mercury. But this is a nonsense. City dwellers all across the globe have seen Mercury with no difficulty at all (when well placed under a clear and transparent sky). I lived in a city for nearly 40 years and enjoyed seeing the planet on many occasions. All one needs to do is to get away from horizon obstructions.

Before moving on to our sister planet, Venus, I would like to bring to your attention another very curious rumor: that Nicolaus Copernicus (1473–1543) is said to have lamented just before he died that he had never been able to see Mercury. To explain the lament, some theorized that the culprit was fog, which was commonly prevalent in the river Vistula, near Thorn, where Copernicus lived.

Others denied the whole story, claiming the belief arose from a misconception. To answer the question, perhaps for the soothing of his own soul, William F. Denning of Bristol, England, studied the naked-eye appearance of Mercury under various atmospheric conditions from his home in Bristol. He found that the planet's brightness could be "impaired by smoke from the city rolling along the western horizon," and "sufficient fog near the horizon" appeared to be enough to obscure the faint object. In an 1893 issue of *The Observatory*, Denning drew a wise and elegant conclusion about the matter:

> Without, however, discussing the evidence on either side, there is no question that the statement has

June 3, 2007 7:45 pm

encouraged observation rather than otherwise. Amateurs find a certain attraction in trying to accomplish a feat which baffled the Genius who "restored law and order to the solar system," and demonstrated the truth of that theory which so simply and effectively explains the motions of the celestial bodies. It would therefore be a pity to disillusion us in regard to the tradition relating to Copernicus and Mercury, for besides providing us with an impressionable incident it adds a peculiar charm to our own observations.

Venus

Roman goddess of love and beauty, high priestess of the twilight sky, Venus has been eternally adored. To the Egyptians, Venus was "the celestial bird of morn and eventide." The Indians called her "the brilliant." And the Arabs saw it as "Zorah, the splendor of the heavens." The ancient Romans dedicated their largest temple to her honor. The Mayan of old identified the "star" as their great god, Quetzalcoatl; they adorned their temples with glyphs of the planet and decoded the planet's motions in the sky and used them to predict the future.

Throughout the ages, writers have immortalized her in their poems and songs. Venus, for instance, is the only planet Homer describes in *The Iliad*, writing, "Hesperus [Venus], most beautiful of stars in the sky." In his acclaimed painting, The White House at Night, Van Gogh accurately depicts the twilight appearance of Venus in June 1890, as it shined over a small white house in the small town of Auvers-sur-Oise, northwest of Paris; Van Gogh, of course, used his artistic license to show Venus as an extraordinarily brilliant yellow star surrounded by an exaggerated dashed halo of yellow light, making it appear more like a celestial crop circle than a glowing

planet. And who of my generation will forget the Shocking Blue's hit single "Venus"? The song began with these charming lyrics:

A Goddess on the mountain top / Burning like a silver flame / The summit of beauty and love / And Venus was her name. Wa!

Today the planet still stuns stargazers of all ages with her vestal radiance. No object other than the Sun or Moon shines as brightly as Venus. When dusk falls over the landscape, or whenever the rose flush of dawn first appears, Venus may shine forth as a stellar beacon bright enough to attract the attention of the most casual onlooker. The image above, for instance, shows Venus' radiance as it is vying for attention against an erupting Kilauea volcano in Hawaii. Not even the daytime sky can completely extinguish the stellar form of Venus, the color of which morphs from opalescent during the day, to vestal white in the early twilight, to silvery white during mid-twilight, and to yellow-white when seen low in a darkening sky.

Roman emperor Cicero (106–43 BC) named Venus as the "morning star," *Lucifer*, which means "bringer of light" (the Greek equivalent is *Phosphorus*). Cicero borrowed the title from the prophet Isaiah (Chapter 14; verses 12–15), who uses the motion of Venus as a metaphor for Satan's fall: "How art thou fallen from Heaven, O Lucifer, son of morning! how art thou cut down to the ground. For thou hast said in thine heart . . . 'I will ascend into heaven . . . ' Yet thou shalt be brought down to hell . . . " Isaiah's account is an extraordinary description of Venus as it rises to heavenly prominence in the east after inferior conjunction . . . only to fall from grace months later, when it plunges beneath the eastern horizon and into the Sun's fiery realm.

Interestingly enough, Lucifer is, ironically, one *hell* of a word to describe the planet's satanic surface. When I was growing up in the late 1950s and early 1960s, astronomers had no way to see the planet's surface, which was perpetually shrouded in dense clouds. But radio

The rise and fall of *Lucifer*

First appearance in east

Greatest elongation

Last appearance in east

telescopes did reveal Venus' surface temperature to exceed 500 °F. Still, it was common to see Venus depicted in books, art, and movies, as a prehistoric world replete with dinosaurs, jungle swamps, and vast roiling oceans – an idea perhaps first put forth by Svante Arrehenius, who, in his 1918 book *The Destinies of the Stars* George Putnam's sons; New York), described Venus as a humid world, covered in swamps and jungles: "We must therefore conclude that everything on Venus is dripping wet."

Meet Venus

Maximum brightness:	−4.6 (−4.4, during average opposition)
Diameter:	12,100 km (7,520 miles)
Apparent diameter (arcseconds):	66.0 (max.); 9.7 (min.)
Average relative density (water = 1):	5.24
Weight; if 100 lb on Earth:	88 lb
Mean distance (from Sun):	108,000,000 km (67,000,000 miles)
Mean orbital velocity	35 km (22 miles) per second
Surface temperature (daytime):	465 °C (870 °F)
Period of revolution:	225 Earth days
Length of day:	243 Earth days (retrograde)
Orbital inclination:	3.4°
Axial tilt:	178°
Number of moons:	0
Rings:	No

How wrong he was! Indeed, NASA spacecraft have shown the surface of Venus to be a hellish wasteland – a world dominated by *thousands* of surface-shaping volcanoes and vast barren flatlands . . . without a trace of water!

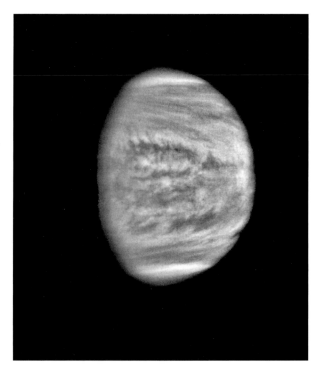

surface temperature is higher. In fact, Venus is the hottest planet in our Solar System.

You couldn't stand on the planet's surface, which has a crushing atmospheric pressure 92 times that of Earth's – the densest in the Solar System. Even if you could move, you'd find it hard to escape the day's heat, which lasts 243 of our Earth days. Equally strange, Venus' polar axis is flopped nearly 180° – perhaps a result of a planetary impact long ago in the planet's history. As a result, Venus is the only planet in the Solar System to rotate backwards. The image at left shows the cloud-draped world as imaged by the Galileo spacecraft and its surface as imaged by radar aboard NASA's Magellan spacecraft (below).

As with Mercury, the morning and evening apparitions of Venus were first believed to be the movements of two different worlds. Thus Venus as the evening star became known as *Hesperus*, or *Vesper*. William Wordsworth praised Vesper in his 1838 poem, "To the Planet Venus":

> What strong allurement draws, what spirit guides,
> Thee, Vesper! brightening still, as if the nearer
> Though com'st to man's abode the spot grew dearer
> Night after night? True is it Nature hides
> Her treasures less and less.

Lightning strikes are common, and, unlike little Mercury, Venus is an Earth-sized world shrouded in dense sulfuric acid clouds. The planet's atmosphere is 96.5 percent carbon dioxide, with a runaway greenhouse effect that raises temperatures to a devilishly hot 465 °C (870 °F); although Venus is farther from the Sun than Mercury, its

Vesper is also the Latin name for the evening bat, of the family *Vespertillonidae*, which first appears in the twilight, darting up and down in the dusk as it feeds.

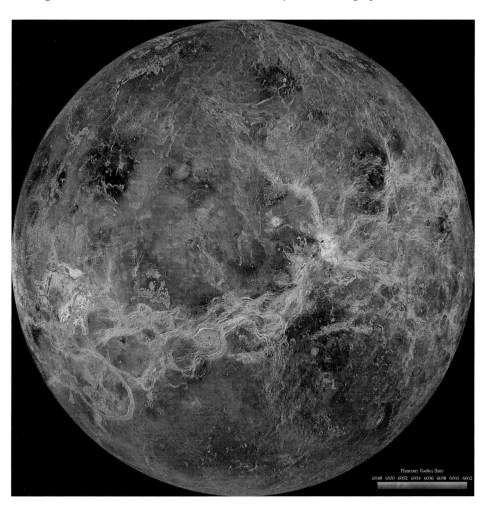

Planetary Radius (km)
6048 6050 6052 6054 6056 6058 6060 6062

Since Venus is farther away from the Sun than the pink-iron planet Mercury (but still inside Earth's orbit), we see Venus move farther away from the Sun than Mercury does in our sky. At maximum elongation, Venus is 47° from the Sun, so its light can be viewed for about an hour after sunset, or before sunrise, against a dark sky.

We first begin to see Venus' angelic light about six weeks after superior conjunction, when it shines as a conspicuous, silvery "star" in the dusk about 30 minutes after sunset. As the nights progress, Venus moves farther and farther away from the Sun's glare, setting later and later each night. It also gradually grows in brightness as the planet sails ever closer to us in its orbit. About seven months after superior conjunction, the planet achieves its greatest western elongation.

Around this time, binoculars will show the planet in a distinct half phase. You'll find the phases of Venus much easier to see than those of Mercury through your binoculars. Venus is not only twice as large as Mercury but has more than six times its surface area — so the planet exhibits a larger area of illumination.

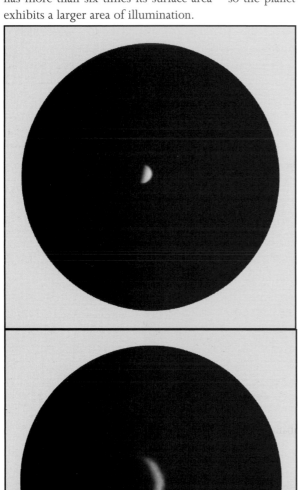

The planet's apparent size continues to increase dramatically after greatest elongation, until the planet reaches inferior conjunction when it's only some 40 million kilometers (25 million miles) distant. The apparent size of Venus at inferior conjunction is about eight times larger than it is at its farthest distance at superior conjunction. And though Venus is only a crescent around the times of inferior conjunction, that's when Venus shines most brilliantly, owing to its nearness.

Because of its great brightness and size, many have claimed to have seen Venus' crescent with their unaided eyes. Take, for instance, this account by Dr. Fred W. Wood, of Alhambra, California, who, in a 1935 *Journal of the Royal Astronomical Society of Canada*, wrote the following account:

> In early January, 1934, I viewed the crescent of Venus as it was near maximum. It was approximately 4:30 p.m. as I opened my garage for my automobile, and as I started to close the same noticed Venus shining through a faint haze or "high" fog as known out here. I was amazed at the clearness of Venus and saw a perfect crescent. I immediately called Mrs. Wood and she confirmed the same. I could easily observe the crescent from any place about the western part of our home lot. I took my 8-power binoculars and verified my naked-eye observation. As it grew dark and in the early twilight the crescent could not be seen except with the telescope.
>
> This is the second time I have seen Venus as a crescent. Once before it was when Venus was near conjunction, about two and a half years ago, under similar atmospheric conditions.

Such sightings are not new. In a 1916 *Journal of the Royal Asiatic Society*, Joseph Oxford quotes the following passage from cuneiform literature as recorded by the early inhabitants of Mesopotamia:

> If on the right horn of Venus a star is visible, you will have
> good crops in the land;
> When upon the right horn of Venus a star is not visible, the land
> will bear many misfortunes.

Years of visual study into this matter has proven to me that, in the least, the planet can be clearly discerned as "out of round", or thinly elongated, when seen by an experienced observer with good eyesight in the twilight. My wife, Donna, who has *excellent* eyesight (20/10), has definitely pointed out the naked-eye crescent to me on more than one occasion, describing the correct orientation and shape, which I confirmed in binoculars. Now consider these words by the poet Shelley, who, in his 1818 poem "Homer's Hymn to Venus," tells how the beauty "Shot forth the light of a soft starlight smile."

Venus is so bright that not only can we see it in the daytime, but it can also cast shadows at night – a fact known to Pliny the Elder. But this requires a completely dark sky far away from city lights. In his 1923 book *Dreams of an Astronomer*, the great French astronomy popularizer, Camille Flammarion says that "This is easily seen either in a dark room or when walking past a wall in the country." On many occasions, I have held a pencil or finger up to a clean white sheet of paper when Venus was at its greatest brightness in a dark sky and have distinctly seen shadows. If the ground is at all rocky, you can also see shadows cast by the stones, giving the dirt an ashen sheen, which is dappled with tiny black shadows.

The planet is also bright enough to create some atmospheric phenomena, including green and blue flashes, many of which I have documented over the years. In fact, the most intense green flash I have ever seen (and the most pure to its name . . . a true flash of light) occurred on May 25, 2002. My wife Donna and I were sitting on a porch in Volcano, Hawaii, just enjoying the night, when Venus, in a dark sky, began to set behind the long slopes of Mauna Loa volcano many miles in the distance. When the bright light of Venus touched the mountain, its white form quickly turned a pure emerald green, which began to shrink; then, suddenly, it flared in magnitude appearing as a tiny green pip of light surrounded by large petals of emerald green. The phenomenon was so intense, that both Donna and I jumped up and screamed simultaneously, "Green!"

Binocular observations of the green flash of Venus are quite common and extremely exciting. Again, the purity of the color is unmistakable and cannot be mistaken for an afterimage, especially since it can occur in a dark sky. But Venus can cause other atmospheric phenomena as well. When its light interacts with water drops and ice crystals, Venus has been known to make ice pillars in the sky – formed when light from the planet reflects off the top surfaces of horizontal plate crystals, which float like leaves in the air – to glitter paths in the sea.

Venus in the daytime

Perhaps the favorite pastime of many observers is using their unaided eyes and binoculars to sight the pearly light of Venus in the daytime without the assistance of a solar eclipse. The best time to search for Venus in the daytime is when the planet is at or near greatest elongation. That's when the planet is farthest from the Sun's glare (about 47°) and is situated in a region of sky that gives maximum contrast . . . if the sky is crystal clear with no contaminants. You'll need to follow the literature or check internet sources, to find out when these elongations will occur. The dates will vary from year to year.

For beginners, it's best to search for Venus in the daytime when it is at greatest western elongation, which occurs when the planet rises before sunrise. This gives you the advantage of being able to follow the planet with your unaided eyes and binoculars up until the moment of sunrise and beyond.

One problem, though, is that waiting for the end of dawn may require at least 30 to 45 minutes of your time. That's why I suggest, as soon as twilight begins, you position Venus so that it just sits atop a familiar landmark, like the edge of your house, or a distant pole or tree. This will give you a terrestrial point of reference in case you decide to walk away from the task for a few minutes to rest. Blocking the point of sunrise and dawn's bright glow with something large like your house has an added benefit: It will shield your eyes from sideways incident light, which reduces contrast. (That's why you involuntarily shield your eyes with your hands when you're searching for something – like your pet cat in a field – under bright sunlight.)

Still, if you look away from Venus, even for a moment when the sky is bright, the planet may be difficult to find again. The spot of Venus is of such low contrast, and the area of sky you have to search is so large (especially with your unaided eyes), that it's really like trying to find the pearly *head* of a pin in a haystack of blue-white light. Ironically, once you lock your sights onto Venus again, the planet appears remarkably obvious. Understanding why is important to the search.

Daytime viewing is completely different than nighttime viewing. Amateur astronomers quickly learn that the best way to see a faint object at night is to place it in the wide periphery of their vision, so that the light falls on the night-sensitive rod cells, which line that part of the eye retina. Daytime observing, however, is completely different. We need to place the object of interest in the extremely narrow central part of the eye's retina, in the macula, or yellow spot, where the tightly packed, detail-sensitive cone cells reside. The macula gives us a sharp field of view measuring only about 2° across. To see Venus in the daylight, then, we need to look *directly* at the planet.

If you lose sight of Venus, sweeping the sky with wild abandon usually does not help. You need to be precise in your visual attack. Plan to look intently at a specific area of sky. If you do not see Venus, move your eye slightly away to a new point, and stare intently again. It takes time, so you must be patient.

Using binoculars will narrow the search area greatly. They will also magnify the planet's image and increase the apparent contrast. Following Venus into the daytime with mounted binoculars in the manner described above is almost a cinch, and you're virtually guaranteed success, as long as you remember to continually move the binoculars to keep pace with Earth's rotation – *and*, most important, you know how to focus your eyes at infinity.

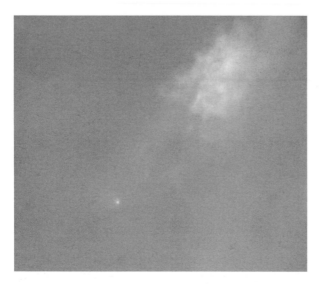

Just as a camera set on autofocus will not be able to achieve critical focus in a fog (the lens just jerks haphazardly in and out of focus), your eyes do not know how to focus on an unseen object in a blank blue sky. Part of the problem is that your mind is at work trying to find the object. If you temporarily lose sight of Venus, do not try to force a focus. What you need to do is relax your eyes and *daydream*.

As silly as it sounds, it works! When a person daydreams, they enter VEM (vapid eye movement) sleep. It is the conscious mind's equivalent of REM (rapid eye movement) sleep. During VEM, our pupils dilate and our eyes focus at infinity. So, if you lose sight of Venus through your binoculars, relax your mind, think of something pleasurable, and let your mind slip into VEM sleep for about five seconds... *then* start your search in earnest. The planet should snap into view once you hit the right spot. Again, think precision attack.

If you're really lucky, a wispy cloud, a milkweed seed, or a high-flying jet, will pass through your field of view. When it does... your eyes will *snap* to an infinity focus, making the task of finding the planet that much easier. In fact, on January 12, 2007, I used some patches of wispy cirrus cloud to help me find Comet C/2006 P1 (McNaught) in broad daylight from the nearly 14,000-foot-high summit of Hawaii's Mauna Kea volcano (see image at left). The comet was shining at roughly −5 magnitude − which is just about a half magnitude brighter than Venus at maximum intensity − only about 10° from the Sun. The image to its below shows how clouds and the Moon can help you to pinpoint the planet in broad daylight.

Indeed, the challenge of spying Venus in the day is also greatly eased whenever the Moon makes a close passage to it. When you see Venus next to the Moon in a midday sky, you'll wonder why the planet is so difficult to see when the Moon is absent. It's simply because you have the Moon not only as a surefire guide, but also at a point of focus.

And you don't even need the Moon or clouds. Throughout history Venus has been spotted in broad daylight by people "on the street" just by looking up and having the planet serendipitously near a distant terrestrial object. Henry White Warren describes one such testimonial in his 1895 book, *Recreations in Astronomy With Directions for Practical Experiments and Telescopic Work* (Harper & Brothers; New York). It describes a sighting made in May 1796, at the high ceremonial given in honor of Napoleon Bonaparte after he defeated the Austrian army at Lodi in Italy:

> Upon repairing to Luxembourg when the Directory was about to give him a fete, [Bonaparte] was much surprised at seeing the multitude paying more attention to the heavens above the palace than to him

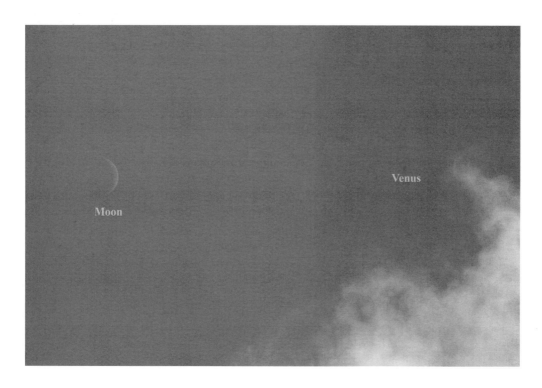

or his brilliant staff. Upon inquiry, he learned that these curious persons were observing with astonishment a star which they supposed to be that of the conqueror of Italy. The emperor himself was not indifferent when his piercing eye caught the clear lustre of Venus smiling upon him at midday.

And, on on March 4, 1865, Sergeant Smith Stimmel, one of Lincoln's bodyguards sighted Venus with the unaided eye on the day of Lincoln's second inaugural. Writing in his 1928 book, *Personal Reminiscences of Abraham Lincoln* (W. H. A. Adams; Minneapolis), Stimmel says, "Shortly after we turned onto Pennsylvania Avenue, west of the Capitol, I noticed the crowd along the street looking intently, and some were pointing to something in the heavens toward the south. I glanced up in that direction, and there in plain view, shining out in all her starlike beauty, was the planet Venus."

Once you become proficient at seeing Venus at greatest elongation in the daytime sky with binoculars and the unaided eye, you might want to start challenging to see the planet as it nears the Sun. Remember, of course, to take every precaution to occult the Sun with some terrestrial object as you look, especially with binoculars!

Transits of Mercury and Venus

WARNING!

To observe a transit of Mercury or Venus, you must use a safe solar filter with your unaided eyes or binoculars. Do not point unfiltered binoculars at the Sun, it could cause permanent eye damage. You can view these events in complete safety by following the advice given on page 5.

The orbits of Mercury and Venus are inclined 7° and 3.4°, respectively, to Earth's orbital plane. Consequently, when the planets pass through inferior conjunction, we commonly see their disks pass well above or below the Sun's, which measures only about 30 arcminutes in apparent diameter. But occasionally the planets do cross, or *transit*, the Sun's face.

Being the smaller of the two worlds, and the most distant from Earth, Mercury is more difficult to detect crossing the Sun than Venus. Mercury's tiny silhouette appears about 200 times smaller than the Sun's disk, or about the size of an average sunspot, so it can be easily mistaken for one. Binoculars are required to see it. The silhouette of Venus, on the other hand, is about six times larger than Mercury's at inferior conjunction, so it is easily detectable with the unaided eye (as through a safe solar filter) when it transits the Sun, appearing as a black disk of "unusual magnitude and of a perfectly circular shape" – a description British astronomer Jeremiah Horrocks made during the first recorded transit of Venus in 1639, and one still held in high regard today.

Transits are fascinating spectacles for the patient and curious binocular observer. All manner of visual phenomena can accompany the silhouettes as they drift across the Sun's face. The larger the binocular the better, but all the phenomena described below have been seen by ordinary handheld binoculars with proper solar filters.

One of the most famous effects, known as the *black-drop effect*, can occur when the planets first enter, or are about to exit, the Sun's disk. After the inferior planet enters the Sun's disk, and just before its following edge separates from the Sun's interior limb, the planet's silhouette may suddenly appear distorted. It may begin to stretch, as if it were stuck to the Sun's limb like a wad of gum to a shoe. As the planet distances itself from the Sun's limb, the stretched ligament begins to thin, until it pinches off from the Sun's limb. Sometimes the appearance is more oblong than linear. For instance, in 1761, William Hirst observed the transit from Madras, India, noting that "At the total immersion of the planet, instead of appearing as

Black-drop effect

truly circular, [it] resembled more the form of a bergamot pear . . . "

What causes the black-drop effect? Irradiation – the same visual effect that makes the earthlit section of a crescent Moon look smaller than the bright crescent surrounding it (the new Moon in the old Moon's arms). When we see an inferior planet necking with the inside of the Sun's limb, irradiation causes the planet's silhouette to appear smaller than expected and a bright object to appear larger when it is seen against a dark background. The black drop occurs when the true limbs suddenly make contact, preventing light escaping from that point.

You can simulate the black drop by choosing a bright background, closing one eye, and bringing together slowly the tips of your thumb and middle finger a few inches from your open eye. But irradiation effects can also be caused by eye movements, spherical aberration in the eye lens, the diffraction of light at the pupil, the way light stimulates the retina, the spreading of photons by rapidly moving air cells, and other optical conditions. So it's a fascinating phenomenon to witness.

Once an inferior planet is completely on the Sun, look for another optical and physiological phenomenon: a bright white ring around its silhouette. While one could argue that the bright white ring around Venus is due to sunlight refracting through its atmosphere, airless Mercury's silhouette shows the same bright aureole. The culprit is eye fatigue. If you stare at a dark object long enough against a bright background, a portion of the eye's retina becomes fatigued, forming a negative (or bright) afterimage. Natural eye movements can then carry the bright afterimage beyond the borders of the dark object, creating the illusion of an aureole.

But Venus' atmosphere does contribute to another remarkable phenomenon just before the planet's trailing

limb fully enters the Sun's disk, or just as leading limb exits the Sun's disk. At these times it's possible, under good definition, to see a bright crescent of light protruding from the Sun's disk – a result of the reflection of the solar light on the atmosphere surrounding the planet. This ring of light was first noticed during the transit of 1761, and has been seen during transits ever since.

Of course, these and other optical effects – colored fringes (caused by atmospheric dispersion) and multiple disks (caused by atmospheric instability) – are largely products of eye magic, or physiological and atmospheric sleights of hand. Still, taking the time to look for these effects with your binoculars makes the otherwise uneventful long passage of these planets across the Sun's face extraordinarily exciting, if not *magical*!

The superior planets

Superior planets, those farther from the Sun than the Earth, can appear anywhere along the ecliptic at any time of night. Their movements are not confined to the twilight regions of the Sun. A superior planet is best seen, however, around the time of opposition, when Earth lies between it and the Sun. That's when the planet is closest to Earth and its face fully illuminated.

While superior planets do not undergo all the dramatic phase changes we see on Mercury and Venus, the closest one, Mars, can be seen in a slight gibbous phase through large binoculars – as much as 89%, like the Moon three days from full. Even Galileo, with his crude instruments (which were inferior to many of today's quality binoculars), believed he could perceive the difference: "[U]nless I am deceiving myself, I believe I have already seen that it is not perfectly round."

But the superior planets do perform a curious dance in the sky around the time of *quadrature* – when the planet lies at an angular distance of 90° east or west of the Sun as seen from Earth. Most of the time, we see the superior planets moving eastward along the ecliptic. Once a superior planet reaches quadrature, however, it appears to slow down, then stop altogether. At this time, the Earth is moving directly towards the planet in its orbit, making it appear to be stationary against the stars.

But as Earth with its greater speed overtakes the superior planet, we see it suddenly appear to reverse its course, moving westward against the stars. Then, as Earth and the superior planet continue in their orbits, the superior planet appears to move forward, when it once again resumes its forward (eastward) march . . . until the process repeats itself again.

Being the planet closest to Earth, Mars shows the greatest and most dramatic retrograde motion. Indeed, early Egyptian stargazers singled Mars out as *sekhed-et-em-khet-ket*, he who moves backwards. Let's look now at this fascinating world.

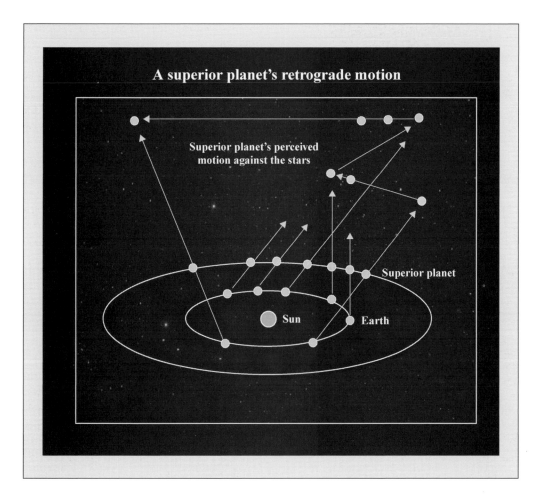

A superior planet's retrograde motion

Superior planet's perceived motion against the stars

Superior planet

Sun

Earth

Mars

More than any other planet, Mars has toyed with man's emotions throughout the ages. The planet's devilish red color is, in part, responsible.

"Unfortunate Mars!" writes the great nineteenth-century French astronomy popularizer Camille Flammarion in his 1904 book, *Astronomy for the Amateur* (Thomas Nelson and Sons; London), "What evil fairy presided at his birth? From antiquity all curses seem to have fallen upon him. He is the god of war and carnage, the protector of armies, the inspirer of hatred among peoples, it is he who pours out the blood of humanity in hecatombs of the nations."

In ancient Babylon, bloody Mars was known as *Nirgal*, the Star of Death. In India, it was *Angakara*, the Burning Coal. In China, it was *Houxing*, and in Japan, *Kasaei* — both meaning Fire Star. In ancient Greece, Mars was *Ares*, the the bloodthirsty god of war. Mars, the name Flammarion refers to, was Roman war god. Interestingly, the soil of Mars does indeed have its blood-red component. When ferrous iron (Fe^{2+}) in Mars' soil combines with oxygen, it turns to ferric iron (Fe^{3+}), or rust — the same chemical reaction that occurs in the hemoglobin molecule.

But unlike the proverbial little red schoolhouse, Mars is not red. Matthew Golombek, project scientist for NASA's Mars Pathfinder mission says "it's actually brownish yellow" — like butterscotch. This is quite interesting, because the origins of the name Mars are still steeped in mystery. Its root may be *mar* or *mas*, which, some linguists argue, are words signifying the generative force. As William Sheehan and I describe in our 2001 book, *Mars: the Lure of the Red Planet* (Prometheus Books; New York), in ancient Roman times Mars was seen as a wheat-colored orb named *Silvanus*, the god of vegetation and fertility. How can the red planet also be a wheat or butterscotch world?

One of the most subtle, yet fantastic observations a binocular observer can make is to document the color of

Mars oppositions

Mars during every apparition. It changes! Depending on orbital, physiological, optical, and atmospheric circumstances (on both Earth and Mars), the planet can appear as any variety of warm shade – from red to pink to orange to gold to yellow.

Mars rounds the Sun in an orbit that's more elliptical than Earth's. As a consequence, the planet is 50 times brighter when nearest Earth than when farthest away. So as Mars grows in prominence against the dark sky, the contrast between it and the night sky also increases, which can explain why Mars can appear to change from a warm hue to a cooler (whiter) shade. That's because the color of Mars as seen with the unaided eyes or binoculars is not extremely saturated. The higher the contrast between the planet and sky, the greater the desaturation of color.

Of course, any celestial object seen close to the horizon will be reddened. When a warm-hued planet like Mars is near rising or setting, atmospheric extinction will dim the planet's light, decrease the contrast between the planet and the sky, and redden its appearance, thus increasing its apparent saturation! The effect is magnified when pollutants infiltrate the air; the more pollutants, the redder the object will appear. As the planet rises, contrast increases, saturation decreases, and the planet yellows.

Consider this description of Mars rising, penned by the great Mars observer Percival Lowell in his 1895 book, *Mars* (Houghton Mifflin Company; Boston):

> . . . a great red star that rises at sunset through the haze about the eastern horizon, and then, mounting higher with the deepening night, blazes forth against the dark background of space with a splendor that outshines Sirius and rivals the giant Jupiter himself.

This is why it's important to make any color observations of Mars when it is highest in the sky. For the same reasons discussed above, you may also notice that the red planet's hue will appear coolest when closest (and therefore brightest) than when it is seen at the more distant parts of its orbit, when it is dimmer. Because Mars travels in a more elliptical orbit than Earth's it draws nearer to us at some apparitions than at others. When closest to Earth (perihelion), the planet is about 56 million kilometers (35 million miles); when farthest away (aphelion), it is about 100 million kilometers (62 million miles) distant.

During perihelic apparitions,[2] when Mars is closest to the Sun in its orbit, the planet appears about 70 times smaller than does the Moon at opposition as viewed from Earth. During Mars' aphelic apparitions, when the planet is farthest from the Sun, the world appears twice smaller still. When Mars makes it first and last appearances during an apparition – well before or after opposition – its disk can appear 300 times smaller than the Moon's! So, as the apparition of Mars waxes and wanes, so too does the planet's apparent size!

During its best perihelic oppositions, Mars can shine at nearly −3 magnitude. At its best aphelic oppositions, the planet shines at only −0.8. The difference can be visually quite striking. Indeed, in 1719, the planet appeared more brightly than it had in 300 years. As the planet neared opposition and that intense blood-red beacon rose above the eastern horizon after sunset, panic ensued among the

---

[2] An apparition is the several-month period between successive conjunctions of a planet with the Sun (as viewed from Earth), during which time the planet is favorably placed for observation.

Mars at its closest, near opposition

Mars when farther away, before opposition

masses. Some thought it to be one of the bloody swords of heaven (a comet), while others thought it was a satanic reincarnation of the Star of Bethlehem.

Speaking of panic, Percival Lowell did more than simply study Mars. He literally brought the Red Planet to life. Mars is the only Earth-like world in our Solar System with surface features clearly visible through a telescope – including polar caps, dark green surface markings, ocher deserts, and atmospheric clouds and vapors. After years of studying the red planet telescopically, after witnessing its dark markings change in concert with its growing or shrinking polar caps, and after documenting the existence of dozens of canals, Lowell came to an astounding con-

clusion: Mars was populated by a dying race of intelligent beings, who had built canals to transport water from the planet's poles to its thirsty desert inhabitants.

In his book *Mars*, Lowell assesses the situation:

That Mars seems to be inhabited is not the last, but the first word on the subject. More important than the mere fact of the existence of living beings there, is the question of what they may be like. Whether we ourselves shall live to learn this cannot, of course, be foretold. One thing, however, we can do, and that speedily: look at things from a standpoint raised above our local point of view; free our minds at least

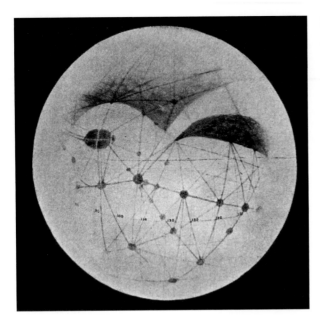

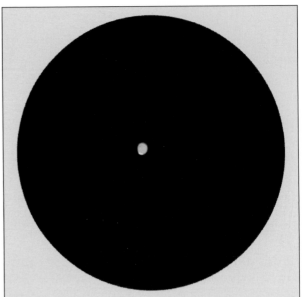

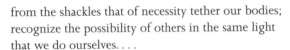

from the shackles that of necessity tether our bodies; recognize the possibility of others in the same light that we do ourselves. . . .

If astronomy teaches us anything, it teaches that man is but a detail in the evolution of the universe . . . He learns that, though he will probably never find his double anywhere, he is destined to discover any number of cousins scattered through space.

Although Lowell was wrong in his thinking, he inspired generations of science-fiction writers to carry his torch. Books and movies abound about life on the planet and alien invaders from Mars. But Lowell's curiosity also spilled over into the minds of scientists, like Carl Sagan, who wondered, at least, about the possibility of microbial life on Mars, and other specialists who helped design an armada of spacecraft missions to the planet. To this day, the scientific results continue to surprise us.

Mars, the fourth rock from the Sun, is a world half the size of ours. At first blush, it appears to be a lifeless, waterless, barren world with many spectacular geological features. Mars is home to Olympus Mons, the largest volcano in the Solar System, soaring to a height more than twice that of Mount Everest. It has its own Grand Canyon – called Valles Marinaris – which is so vast it could extend across the length of the United States! It also has polar caps large enough at certain Martian seasons to be seen by a keen observer through large mounted binoculars, especially with the aid of a blue filter. Expert binocular observer Phil Harrington of Long Island, New York, and author of *Touring the Universe Through Binoculars* (John Wiley and Sons, Inc.; New York, 1990), says that during the favorable Mars apparition of 2003, he saw "some evidence" of a Martian polar cap through a friend's 30 × 80 binoculars. The photo illustration at above right shows how Mars may appear during a perihelic opposition as seen through large mounted binoculars with the aid of a blue filter.

Meet Mars	
Maximum brightness:	−2.9 (−2.0, during average oppositions)
Diameter:	6,790 km (4,220 miles)
Apparent diameter (arcseconds):	25.1 (max.); 3.5 (min.)
Average relative density (water = 1):	3.9
Weight; if 100 lb on Earth:	39 lb
Mean distance (from Sun):	228,000,000 km (142,000,000 miles)
Mean orbital velocity	24 km (15 miles) per second
Surface temperature (daytime):	−30 to −40 °C (−20 to −40 °F)
Period of revolution:	687 Earth days (~2 years)
Length of day:	24.66 hours
Orbital inclination:	1.9°
Axial tilt:	25.2°
Number of moons:	2
Rings:	No

But the barrenness may be a thinly disguised veil. Evidence is now mounting that the climate of Mars today was probably unlike that of its past, when it was warmer. Oceans of water may have covered large swaths of the planet's surface, and ancient rivers appear to have cut long channels, valleys, and gullies on the planet's surface. Spacecraft have found water-ice hiding beneath the planet's south pole, and evidence is mounting that water may still exist on Mars, as a permafrost trapped in cracks and pores in in subsurface rock. Equally tantalizing, and highly debated, is the possibility that fossil microbial life exists in a meteorite that fell to Earth from Mars.

If true, Mars may have had life before life on Earth had a chance to take hold. It may also mean that there is a far-reaching possibility that the first life on Earth was transported here from Mars. No matter how you slice it, Mars has at least three ingredients necessary for life: liquid water, the necessary chemical elements to build life (carbon, hydrogen, oxygen, and nitrogen), and sunlight and geothermal heat as viable energy sources.

Mars has two tiny moons, Phobos (Panic) and Deimos (Fear). These are irregularly shaped bodies and may be captured asteroids. Phobos is about 27 km (17 miles) in diameter and shines around magnitude 11 when brightest. Deimos is about half as wide as Phobos and one magnitude fainter. These magnitudes are within range of large-binocular users. But the problem is the intense glare of Mars. Without Mars, both would be easy enough to see.

Also, when Mars reaches opposition during a perihelic apparition, a considerable amount of dust can be suspended in the Martian atmosphere. Sunlight reflecting off this dust can make the planet appear brighter, and therefore whiter, to the naked eye than normal; for a butterscotch planet this means a color shift toward yellow. This was the case during the great July 2001 dust storm on Mars. As reported in the December *Circular of the British Astronomical Association* Mars Section for that year, the color of the planet "was more yellow than orange and even to the naked eye the colour was noticeably different." Likewise, during the dusty perihelic opposition of 2003, I saw Mars distinctly yellow as seen with the unaided eye and through handheld binoculars.

When Mars is at its brightest (magnitude −2.9), it can be, and has been seen in the daylight with the unaided eye and binoculars; the former being, of course, the more difficult observation. I observed the planet in daylight during the 2003 perihelic apparition. I first lined up the planet with some familiar terrestrial landmarks in the west as the dawn brightened in the east. I then alternated between binocular and naked-eye views until the Sun rose.

One of the earliest daytime observations that I'm aware of was made by Charles Frederick Juritz of Capetown, South Africa. As he reports in a 1910 issue of *The Observatory*, on October 9, 1910, it occurred to him that as

> Mars had been approaching Venus so nearly in brilliancy *after* sunset, the former planet too should be visible to the unaided eye *before* sunset . . . Unfortunately there was but little time that afternoon for making observations, as it wanted less than half an hour to sunset. Mars was, however, so bright that it did not take long to discover with the naked eye, and the planet was thus seen quite distinctly by myself and by three friends, whose attention I drew to the occurrence, from ten to fifteen minutes before sunset; the Sun sets, I may here say, with a clear sea horizon. I at once verified the naked-eye observation by bringing a $3\frac{1}{2}$-inch telescope to bear on the planet, by which means it was also observed by my friends.

Jupiter

Our Solar System's largest planet, Jupiter (see photo below), is named for the Roman king of the gods (Zeus, to the ancient Greeks). Isn't it curious, though, that mighty Jupiter would be named for only the *second* brightest planet. The ancients had no way of knowing that distant Jupiter was an enormous ball of gas and liquid 1,000 times larger than Earth. They couldn't fathom that it would take 11 Earths, lined side by side, to fit across giant Jupiter's equator. Nor did they know that Jupiter lies seven times more distant from us than Venus, or that it shines so brightly because of sunlight shining off of its vast surface area. Yet Jupiter was associated with the great

god *Marduk*, the ruler of the late Babylonian pantheon, and with the great god *Amun* of Egypt.

Yet, to the unaided eye, Venus alone dominates the planets in brilliance. Why then wasn't the name Jupiter bestowed upon Venus? St. Augustine (354–430 CE) asked the same question in the seventh volume of his book, *The City of God*; the following is from Marcus Dod's 1872 translation (T. T. Clark; Edinburgh):

> But since they call Jupiter king of all, who will not laugh to see his star so far surpassed in brilliancy by the star of Venus?... They answer that it only appears so because it is higher up and much farther away from the earth. If, therefore, its greater dignity has deserved a higher place, why is Saturn higher in the heavens that Jupiter?

Perhaps the answer lies in Jupiter's rich golden hue, and the fact that, unlike Venus, it is the ruler most high in the heavens at night.[3] As we see in the *Iliad*, Homer dresses Zeus in gold; he seats Zeus on a golden throne, and looks down on all of human's activities from his lofty perch. Unlike Venus, Jupiter does not hide in the skirts of the Sun. Jupiter has the power to roam the night, to lead the course of the starry heavens across the skies without being "pulled down" by the Sun. "I am mightiest of all," Zeus says "Make trial that you may know. Fasten a rope of gold to heaven and lay hold, every god and goddess. You could not drag down Zeus. But if I wished to drag you down, then I would."

The irony is that Jupiter is so massive that it can pull down any Solar System object that dares test its gravitational strength (except the Sun) – as was the case when comet D/1993F2 (Shoemaker–Levy 9) ventured too close to the giant planet and could not escape its pull. Jupiter's gravity shattered the comet into 21 fragments that came crashing down on Jupiter's atmosphere; the largest fragments detonated with tremendous explosions, scattering Earth-sized clouds of debris across Jupiter's cloud tops; these were visible in the smallest of backyard telescopes, and perhaps even large powerful binoculars.

Like Venus, Jupiter is bright enough to be seen in the day with the unaided eye. Just about any pair of binoculars will show Jupiter as a tiny disk. Careful scrutiny, especially in the dawn or dusk, however, with binoculars that magnify greater than 10×, and certainly those that magnify 20×, will show one or two dusky equatorial belts paralleling the planet's equator. "Even my cheap 12 × 50 Bresser binoculars that I got on eBay for about $20," Phil Harrington says, "showed me them in the summer of 2008." These dark features are separated by equally striking white zones. All are clouds (mostly hydrogen and helium) stretched out into long bands by Jupiter's rapid

Meet Jupiter	
Maximum brightness:	−2.9 (−2.7, during average oppositions)
Diameter:	142,984 km (88,846 miles)
Apparent diameter (arcseconds):	50.1 (max.); 29.8 (min.)
Average relative density (water = 1):	1.3
Weight; if 100 lb on Earth:	240 lb
Mean distance (from Sun):	779,000,000 km (484,000,000 miles)
Mean orbital velocity:	13 km (8 miles) per second
Cloud-top temperature:	−145 °C (−230 °F)
Period of revolution:	4,333 Earth days (~12 years)
Length of day:	9.93 hours
Orbital inclination:	1.3°
Axial tilt:	3.1°
Number of moons:	63; 16 of which are ~10 km (6 miles) wide
Rings:	Yes

rotation; Jupiter is the fastest turning planet in the Solar System, its surface clouds spinning once on its axis about every 10 hours. Because of its rapid spin, Jupiter also appears significantly oblate, being about 7 percent wider at the equator than at the poles.

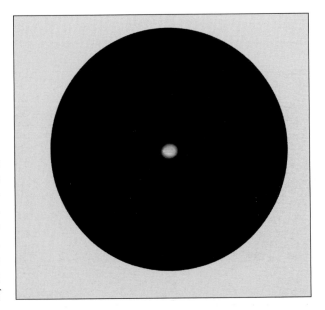

Interestingly, you'll find as little difference between the size of Jupiter and the Sun, as there is between Jupiter and the Earth. No one knows if the planet has a solid surface. But it probably has a rocky core (with layers of metals and ices) some 14 to 18 times more massive than Earth's. That core – a remnant of the initial compression of dust and gas that helped to form the planet 4.6 billion years ago – is still radiating heat, about twice as much as it receives

[3] Jupiter does appear yellow-gold to the unaided eye when it is seen against a dark sky, but I suggest you look at it, especially through binoculars, when it is bright against the fading dusk or growing dawn. The planet then appears as a delicious cream color with a flush of pink. This color is especially apparent when Jupiter is seen next to the "silver fox" (Venus).

from the Sun. In fact, had Jupiter been about 75 percent more massive, it might have become a Sun.

Aside from its belts and zones, Jupiter's most celebrated surface detail is its Great Red Spot, a reddish-brown oval of swirling gas in the planet's southern hemisphere, which has persisted arguably since the invention of the telescope some 400 years ago (see image above). Spacecraft have shown the feature to be highly reminiscent of a hurricane on Earth – though at its widest extent, the Great Red Spot could consume three Earths! The spot rotates counterclockwise (so it's an anticyclone) with a period of about 6 days. No one is exactly sure why it is red. Some have speculated that certain compounds of sulfur and phosphorus in ammonia crystals are responsible for its hue.

In the early 1970s, the spot was so brick red and intense, that I noted it was the most obvious feature

on the planet as seen through a 3-inch (8-cm) finderscope at low power, which means that it could have been detected in one of today's large mounted binoculars. But the Great Red Spot has not always been red, nor has it always been that intense. Ever since the feature became generally known and followed in 1878, the spot has waxed and waned from view, sometimes to near invisibility. The most intense periods of color occurred in the early 1880s, mid 1930s, and early 1970s. So it is difficult to predict exactly when, if ever, the Great Red Spot will return to prominence.

Seven-power binoculars and larger will show the planet's four brightest moons, which are seen in the NASA image below. Galileo first spotted these tiny worlds in 1610, giving them the honor of being the first astronomical objects discovered through a telescope. He named them (in order of increasing distance from the planet) as follows: Io (I), Europa (II), Ganymede (III), and Callisto (IV).

With a diameter of 5,270 km (3,270 miles) Ganymede is the the largest, exceeding Mercury in size and being not much smaller than Mars. Callisto ranks second at 4,800 km (3,000 miles); it is slightly smaller than Mercury. Io is next, measuring 3,635 km (1,940 miles) across. And Europa is the smallest at 3,130 km (1,945 miles). And while Jupiter has more than 60 known orbiting satellites (many of them perhaps captured asteroids in temporary orbits), at least 16 of them measure a respectable 10 km (6 miles) wide. So giant Jupiter really is like a little Sun with its own family of orbiting worlds.

Each of the Galilean satellites is a marvel unto itself. Io is the most volcanically active world in the Solar System, resurfacing itself every year with a thin veneer of sulfurous

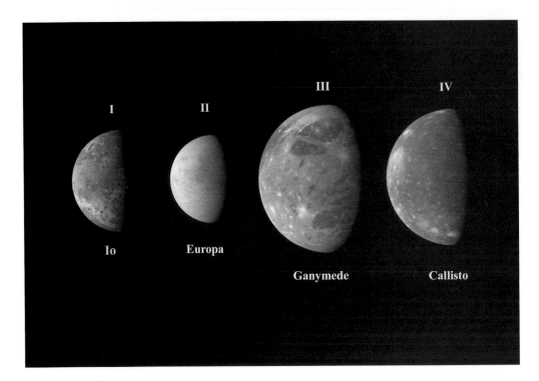

June 5, 2007   8 pm HST

volcanic products. Europa is the smoothest object in the Solar System. Most of its surface is a flat ice sheet criss-crossed with thousands of fractures or fissures, beneath which may exist liquid seas of briny water, heated by geothermal activity. Occasionally, the fissures act as ice volcanoes, erupting material onto the surface and into space. Ganymede, the largest moon in the Solar System, is a world divided. Half of its surface dates to primordial times, as evidenced by intense fracturing and cratering of its surface. The other half has been acted upon by some mysterious tectonic forces, resulting in long paral-lel furrows and grooved terrain. And Callisto is the most cratered body known in the Solar System; its surface being hardly modified since it formed 4.6 billion years ago.

And what a joy it is to see these mysterious worlds through your binoculars, performing slow but precise waltzes around the glowing orb of Jupiter, shifting their positions night after night, in a celestial shell game. Some-times they gather on one side of Jupiter, like kids at play. At other times they appear on both sides of the planet, strung out like a glittering necklace. On other nights, only one or two of the moons may be visible, or per-haps none at all! But they always return, always dance. Like the young girl Karen in the 1845 Hans Christian Andersen fairy tale, *The Red Shoes*, the Jovian moons cannot stop dancing. You can see them hour after hour, night after night, year after year, swinging around the planet to some unheard rhythm . . . but one we know, nevertheless, that governs all the worlds in the silence of space.

"Probably much of the impressiveness of the specta-cle," writes Garrett Serviss in his 1923 book, *Astronomy with an Opera-Glass* (D. Appelton and Company; New York), "is owing to the knowledge that those little points of light . . . are actually, at every instant, under the govern-ment of their giant neighbor and master, and that as we look upon them, obediently making their circuits about

him, never venturing beyond a certain distance away, we behold a type of that [gravitational] mastery to which our own little planet is subject as it revolves around its still greater ruler, the sun, to whose control even Jupiter in his turn must submit."

To know which moons are which, you will need either to purchase astronomy software (such as Starry Night®) or subscribe to an astronomy magazine. Most popular astronomy magazines provide you each month with a night-to-night plot of the changing positions of Jupiter's moons.

Shining at magnitude 4.6, Ganymede is the brightest Galilean moon, and it has been well documented to be seen with the unaided eye. Of course, if it weren't for the intensity of neighboring Jupiter, seeing the tiny little moon would not be difficult at all, even from the suburbs. But we do have to contend with Jupiter's light.

Brightnesses of the Galilean satellites	
(I) Io	5.0
(II) Europa	5.3
(III) Ganymede	4.6
(IV) Callisto	5.6

To see Ganymede, if not Europa (Io is too close to the planet to see; Callisto may be a tad too faint), with the unaided eyes; one needs excellent vision and few aberrations in the eye. In *Light and Color in the Outdoors* (Springer-Verlag; New York, 1993), the late Marcel Minnaert notes that exceptionally good observers can "see two of the satellites of Jupiter, though only at dusk when the stars of the first and second magnitude are beginning to appear."

It's true. On September 8, 1982, I was in Hawaii, pointing out Jupiter in the dusk to a local youth group. I had just finished describing the four Galilean moons, when one of children shouted, "I see one!" Another child confirmed the sighting, exclaiming, "Yes, it's to the upper right. And there's another one to the left." I was skeptical, at first, thinking they were joking me, but the kids kept insisting it was true. So I took the time to look myself and soon saw one object clearly separated from Jupiter to the upper right! The object to the left appeared only as a suspicious bump. Later, I confirmed the positions of Ganymede and Europa through binoculars.

As reported in an 1860 *Monthly Notices of the Royal Astronomical Society* (volume 20, page 212), Captain Boyd, commander of the H.M.S. *Ajax*, communicated to the Astronomer Royal that he, Lieutenant Grant, and other officers aboard his ship, had seen some of Jupiter's satellites with their unaided eyes on January 15. The ship was then in Kingstown Harbour, Dublin, and the weather during the observation was "unusually clear." The satellites and Jupiter were in the following order: 3,2,1, Jupiter, 4.

It's reported that "Capt. Boyd, and perhaps others of the officers, saw 3 and 2 with the naked eye; Lieut. Grant saw 3 only." They confirmed their naked-eye sightings with ship telescopes.

Another wonderful sighting occurred on the night of April 3, 1874, when William F. Denning of Bristol saw Ganymede and Callisto "unmistakeably." The moons were at their greatest elongations from the planet and the night was exceptionally dark. He attributes his success to cutting off the "planet's marginal rays"; the moons were not visible otherwise. He saw them "steadily and separately several times."

Saturn

Lord of the Rings, Saturn is the second largest planet in the Solar System, and the last of the naked-eye planets visible to the ancients. While Jupiter takes 12 years to complete one orbit around the Sun, Saturn takes 29.5 years. So its pace across the night sky is exceedingly slow, progressing eastward by only a little more than about a fist held at arm's length each year on average. Only when the planet is close to a very bright star can we perceive it moving ever so slightly in a matter of days. The images below show Mars and Saturn on June 20, 2008 (left) and

June 23, 2008 (right), near the star Regulus in the Sickle of Leo the Lion. Note how nearly imperceptive the movement of Saturn is in three days, compared to that of closer, and faster-moving, Mars.

Like Jupiter, Saturn is a gas-giant world with an equatorial diameter 10 times greater than Earth's. But because of its greater distance, it shines more dimly, being about as bright, on average, as a 1st-magnitude star, shining with an unfaltering straw-colored light. Perhaps it is due to this color that it was known to the ancient Babylonians as *Ninib* (an agricultural god) and to the Greeks as *Cronus*, the Protector and Sower of the Seed.

Likewise, Saturn was the Roman god of agriculture and harvest. Saturn was said to have ruled over humanity during the Golden Age – a mythical time long ago when life on Earth was a utopia – pure and immortal. In memory of this time, people celebrated the festival of *Saturnalia*, a day event (December 17) that later stretched out over seven days around the time of the winter solstice. Some believe that this festival of gaiety, role reversal, and moral slackness, gave birth to today's festive and global *Carnival* – the most elaborate of which is the New Orleans *Mardis Gras* (Fat Tuesday).

Meet Saturn

Maximum brightness:	+0.4 (+0.7, during average oppositions)
Diameter:	120,540 km (74,900 miles)
Apparent diameter (arcseconds):	20.1 (max.); 14.5 (min.)
Average relative density (water = 1):	0.69
Weight; if 100 lb on Earth:	107 lb
Mean distance (from Sun):	1,427,000,000 km (887,000,000 miles)
Mean orbital velocity:	10 km (6 miles) per second
Cloud-top temperature:	−175 °C (−285 °F)
Period of revolution:	10,759 Earth days (~29.5 years)
Length of day:	10.65 hours
Orbital inclination:	2.5°
Axial tilt:	26.7°
Number of moons:	60; 25 of which are ~10 km (6 miles) wide
Rings:	Yes

Part of Saturn's brightness is due to its glorious rings, arguably the most beautiful telescopic sight in the heavens. While the ball of Saturn measures about 120,000 km (75,000 miles) across, its rings extend to 250,000 km (155,000 miles); they also reflect more light than the planet, when fully open. Under the most favorable circumstance, Saturn, with its fully open rings, can shine at magnitude −0.3; Without them, Saturn's brightness would drop to about magnitude +0.8.

Saturn is not the only planet in our Solar System with rings; Jupiter, Uranus, and Neptune have them as well. But Saturn's rings are unrivaled. None of the other planetary ring systems can be seen from a person's backyard through even the smallest of telescopes, especially at powers of 60× or more. (I will soon discuss if the rings can be seen in binoculars.) Those of Jupiter, Uranus, and Neptune can only be seen in images taken by the world's most powerful telescopes on Earth and space, and by spacecraft.

Saturn's rings are comprised of boulder-sized (and smaller) chunks of dusty water-ice that gently collide with each other as they orbit. But the particles are not uniformly distributed; Cassini spacecraft data show them to clump together to form elongated, curved aggregates, continually forming and dispersing. The space between the clumps is mostly empty.

The rings totally circle the planet. But, owing to Saturn's great axial tilt (~27°), we see the rings undergo a series of dramatic changes as the planet rounds the Sun. Sometimes we view them from above; at other times from below. Twice each orbit, we also get to see them turn edge on, when they virtually disappear. Due to the eccentricity of Saturn's orbit, these ring-plane crossings occur at alternate intervals of roughly 13 years 9 months and 15 years 9 months.

When the rings are *nearly* edge on (which they are for most of the apparition), they appear as a thin thread of ghostly light. Seeing them at all around this time is a marvel: Saturn's rings are only about 10 meters (30 feet) thick, and we're looking at them from a distance of nearly a billion miles. Yet, when the rings are just a few degrees open, a keen-eyed observer under dark skies can spot them through large, mounted binoculars that can magnify at least 20×. Saturn's last edge-on apparition occurred in 2009. Its next edge-on apparition will be in 2025; unfortunately, the planet will be too close to the Sun to view the main event. The next decent time to catch this a ring-plane crossing won't be until 2038–39.

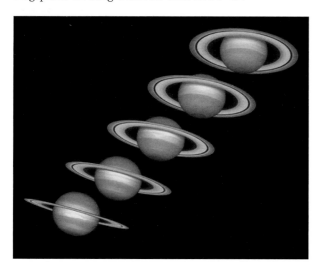

Alas, you won't get a clear view of the rings in handheld binoculars. When the rings are open, you can see that the planet is elliptical in 7 × 35 and 7 × 50 binoculars. But most observers I know, try as they might, have not been able to differentiate the glow of the rings from the glow of the planet with these instruments. What is visible is their combined form, so the planet looks elliptical – the degree to which it does depends on the openness of the rings. If you own a pair of 10× image-stabilized binoculars, or a pair of regular 10 × 50 or 11× 80 binoculars that can be mounted on a tripod, I do suggest that you try separating the rings from the planet. It takes time, patience, and constant vigilance, but you may succeed!

Otherwise, as Phil Harrington told me, 10 × 50s, and even 12 × 50s, "show only an elongated shape when the rings are close to fully open, but that's about it." But, he adds, his 16 × 70 Fujinons "distinctly show dark space between the rings and the planet's disk, which is made even more evident through 20 × 80s."

Theoretically, the dark gap between the ring and planet (when the rings are sufficiently open), should be resolvable at 10×. The problem is cutting down the glare from both the planet and the rings, which causes irradiation effects under high-contrast situations, such as during the darkness of night. You might experiment by using your binoculars as a monocular with an aperture stop. You can also overcome the glare problem by looking for the rings when the planet is in the twilight. Very stable atmospheric conditions are also important. You'll just have to experiment to find out what you can see.

Large mounted binoculars are better for achieving success, especially those that magnify 15× or more. If a keen-eyed observer looks at Saturn in the twilight, when the

contrast between the sky and planet is diminished, he or she may be able to make out the rings' clearly "horned" extensions to the planet. I have done this with Orion 25 × 100 binoculars. Again, the key is to keep at it, and only quit when you feel you've exhausted all possibilities.

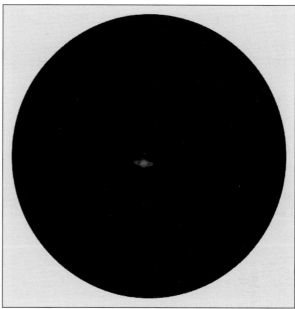

Saturn's ring system is surrounded by more than 60 moons, but only one, Titan, is bright enough to be seen well in moderate-to-large-sized binoculars. Titan is an exciting moon – really a world unto itself. With a diameter of about 5,150 km (3,200 miles), it's larger than Mercury but slightly smaller than Jupiter's Ganymede. But Titan is the only moon in the Solar System with a considerable atmosphere. Early spacecraft images revealed the moon to be veiled in a ubiquitous layer of orange smog, which consists mostly (90%) of nitrogen gas. But in 2005, NASA's Cassini spacecraft not only mapped Titan's surface at wavelengths in which the atmosphere is transparent, but it also launched the Huygens probe, which parachuted to the moon's surface.

The images revealed a "totally alien" world, one that's arguably the most fascinating of all the moons in the Solar System (see the top image on page 96). As Cassini team member Ralph Lorenz said, "It's a giant factory of organic chemicals." Several hundred lakes and seas have been observed, which may contain more hydrocarbon liquid than Earth's oil and gas reserves. Dark dunes with a volume of organics several hundred times larger than Earth's coal reserves have also been found along the moon's equator. Its surface also has rolling Earth-like hills of ice, river channels formed by flowing liquid methane, liquid lakes, and perhaps tar pits.

Titan shines at magnitude 8.3, which is normally bright enough to be easily detectable in handheld binoculars, but it lies so close to the planet as seen from Earth that, even at greatest elongation, it's a challenge to see. I find Titan extremely difficult to detect in 10 × 50 binoculars when the rings are open. This speck of light only gets about as far from Saturn as 6th-magnitude Europa does

from Jupiter. But I did achieve success on June 2, 2008, when Saturn's rings had closed to about 9°. Seeing it, however, required mounting the binoculars and using averted vision to cut down on the glare surrounding the elliptical form of the planet with its unresolved rings. The view below illustrates a clear view of Titan through large mounted binoculars. Titan is the small spot of light to the planet's lower right; the position of the moon and its orientation to the planet will vary over time. When Saturn is most accessible in the evening sky, monthly astronomy magazines usually publish graphs showing the shifting position of Titan (and other bright Saturnian moons) relative to the planet over time.

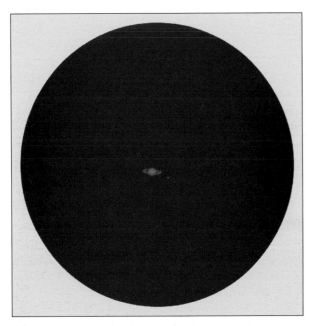

The moon was easy to see in February, 2009, when the rings were only 3° open and "invisible" in the 10 × 50s. It is quite a spectacle in 25 × 100 binoculars. When the rings are edgewise, large-binocular users should also try for Rhea (magnitude 10.0), Tethys (magnitude 10.6),

and Dione (magnitude 10.7) when they are at great elongations from the planet.

Uranus and Neptune

We have now reached the two most distant planets in our new Solar System. The German-born British astronomer William Herschel discovered Uranus telescopically in 1781. He named it *Georgium Sidus* (The Georgian Star), for his patron King George III. Ultimately, however, tradition prevailed, and the new world – the first to be discovered since antiquity – became officially known as Uranus, after the Greek god who first ruled Olympus and the universe.[4] Actually, Herschel was not the first to see it. That honor goes to the Astronomer Royal John Flamsteed, who cataloged the planet six times between 1690 and 1715. But his inferior telescope showed no discernible disk; and since the planet moves so remarkably slowly against the background stars, he simply mistook it for a star, which he named "34 Tauri." Several other astronomers of the day followed suit, independently sighting and recording the "star" without ever suspecting that it was a planet. But Herschel's superior telescope enabled him to see the "star" as an extended object, which he at first mistook for a comet.

Neptune has a similar and equally fascinating history. Charles T. Kowal of Palomar Observatory in Southern California discovered that Galileo had recorded Neptune as a "fixed star" on December 28, 1612. The great Italian observer also saw the "star" again on January 28, 1613. But the observation on this date was extraordinary. Kowal notes that on that evening, Galileo had sketched the positions of two stars: one labeled "A"; and the other, "B". Kowal discovered that star B was indeed Neptune. But he also noticed that Galileo had written a comment in Latin, which he translated:

> After fixed star A, following in the same line, is star B, which I saw in the preceding night, but they then seemed farther apart from one another.

"Not only had Galileo spotted Neptune," Kowal says, "he even noticed that it had moved from night to night!"

In 1795, the French astronomer Joseph-Jérôme de Lalande (1732–1807), or his assistant, twice recorded Neptune while working on a star catalog. But the planet was mistaken for a star, even though the planet had moved slightly against the other fixed stars. No matter, in time astronomers began to notice that the position of Uranus often differed from prediction. Suspecting the gravitational pull of a more distant world was the cause, two young mathematicians – John Couch Adams in England, and Urbain Jean Joseph Leverrier in France – began to

---

[4] All the classical planets are known today by the name of a Roman god, except for Uranus, which is named for a Greek god. The Roman equivalent of Uranus is *Caelus*, from *caelum*, meaning "sky" or "heaven." *Urania* (not Uranus) is the muse of astronomy and geometry.

predict independently (neither one knew of the other) the location of this unseen world – the Solar System's first Planet X.

Both mathematicians sent their final predictions to trusted astronomers in the hopes that they would telescopically spy the new planet. The first to succeed were German astronomers Johann G. Galle and Heinrich L. d'Arrest at Berlin Observatory; on September 23, 1846, they found the planet close to Leverrier's position. Thus, Planet × became the first world whose existence was predicted mathematically. It was named Neptune after the Roman god of the sea – an appropriate reference to this ruler of the "deep."

Today, spacecraft have revealed the wonders of both Uranus and Neptune. Like Jupiter and Saturn, both are gas giants. Measuring 51,118 km (31,763 miles), Uranus beats out Neptune by about 1,500 km (930 miles) to rank as the Solar System's third largest planet. You could line four earths, side by side, across both planets' equators. Uranus and Neptune also have atmospheres that consist primarily of hydrogen, helium, and a small fraction of methane. The methane gives them their interesting colors: Uranus is blue-green, and Neptune is blue (the other Blue Planet).

Each planet also has its own interesting physical characteristics. Uranus, for instance, has the curious distinction of having an axial tilt of 97.8°, meaning that it orbits the Sun nearly on its side (see image below at left). The planet also radiates as much heat into space as it receives from the Sun. And while no one knows for certain about the planet's core, planetary scientists theorize that it's an ice-capped rock, about the size of Earth, surrounded by an ocean of liquid water containing dissolved ammonia.

Meet Uranus	
Maximum brightness:	+5.5 (+5.7, during average oppositions)
Diameter:	51,118 km (31,763 miles)
Apparent diameter (arcseconds):	4.1 (max.); 3.3 (min.)
Average relative density (water = 1):	1.27
Weight; if 100 lb on Earth:	~90 lb
Mean distance (from Sun):	2,872,000,000 km (1,785,000,000 miles)
Mean orbital velocity:	7 km (4 miles) per second
Cloud-top temperature:	−215 °C (−355 °F)
Period of revolution:	30,685 Earth days (~84 years)
Length of day:	17.2 hours
Orbital inclination:	0.8°
Axial tilt:	97.8°
Number of moons:	27
Rings:	Yes

Neptune (see image, below at right) orbits the Sun tilted at an angle of 28.3°. Its frozen methane clouds blow around this world with fantastic speeds, up to 1,100 km (700 miles per second); these are, in fact, the fastest wind speeds in the Solar System. In 1989, the Voyager 2 spacecraft imaged one of the most dramatic cloud features in the outer Solar System: a fantastic hurricane-like storm, called the Great Dark Spot, a feature that resembled Jupiter's Great Red Spot. Hubble Space Telescope images of Neptune in 1994, however, revealed that Neptune's violently swirling storm had vanished. Like Uranus, Neptune's core may be an icy silicate mass surrounded by a compressed liquid. And while both worlds have a number of moons, none are visible to binocular users.

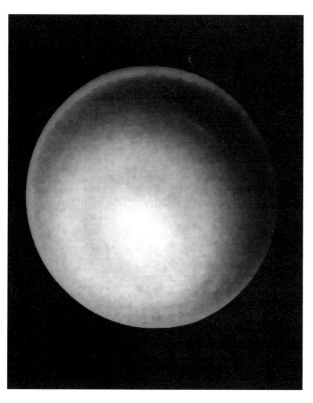

Meet Neptune	
Maximum brightness:	+7.8 (+7.9, during average oppositions)
Diameter:	49,528 km (30,775 miles)
Apparent diameter (arcseconds):	2.4 (max.); 2.2 (min.)
Average relative density (water = 1):	1.64
Weight; if 100 lb on Earth:	~120 lb
Mean distance (from Sun):	4,495,000,000 km (2,793,000,000 miles)
Mean orbital velocity:	5.4 km (3 miles) per second
Cloud-top temperature:	−220 °C (−353 °F)
Period of revolution:	~165 Earth years
Length of day:	16.1 hours
Orbital inclination:	1.8°
Axial tilt:	28.3°
Number of moons:	13
Rings:	Yes

Both Uranus and Neptune are fun to hunt down. But this requires some basic knowledge of the night sky: you must know the zodiacal constellations and how to read a star chart. The night sky has been mapped out in a grid system, which you'll find superimposed on detailed star charts. If you know the coordinates of either planet, you can plot it on a detailed star map just as you would the coordinates of a city on a world map. Astronomy software tools such as Starry Night®, and websites, such as Heavens Above (http://www.heavens-above.com), will provide you with the necessary information and maps, which you can also print out for the binocular view.

Otherwise, most monthly astronomy magazines publish the coordinates of Uranus and Neptune for at least for the middle of the month. The coordinates are given in *right ascension* (R.A.) and *declination* (Dec.) Right ascension is the celestial equivalent of terrestrial longitude. It is measured eastward in hours (h), minutes (m), and seconds (s) – from 0 h to 24 h. Declination is the equivalent to terrestrial latitude. It is measured in degrees (°), minutes (′) and seconds (″) – from 0° to 90° north or south of the celestial equator; an object north of the celestial equator is indicated by a plus (+) sign, an object south of the celestial equator is indicated by a minus (−) sign. All you have to do is look at your star map, find the page that shows the appropriate right ascension and declination and plot it.

When done, you go outside, find that area of sky and point your binoculars in that direction. You're looking for a "star" that is not plotted on your map of that vicinity. The illustrations below, for instance, demonstrate how little Uranus (arrowed in the labeled photo at right can alter the appearance of a star field.

If you have large binoculars that magnify 20× or more, you can see the disks of both planets! Uranus should appear greenish, while Neptune will appear more bluish. The planets' true colors, Phil Harrington, notes, can be seen through 10 × 50 binoculars. But these colors may be highly subjective. On some nights, through 25 × 100 binoculars, I've seen Uranus looking more yellow than green, and Neptune looking more green than blue – perhaps due to the elimination of some shorter wavelengths of light by volcanic pollutants in the air.

Uranus was too faint and obscure to attract the attention of ancient stargazers; besides, what deity would deign appear so diminutive among its heavenly hosts? But we have the advantage of knowing where to look. Today, seeing Uranus with the unaided eye is not an extreme challenge. But it's not easy either. The biggest problem is separating it from nearby stars.

The fact is, an average person with average eyesight under a dark sky can see to magnitude 6 and fainter

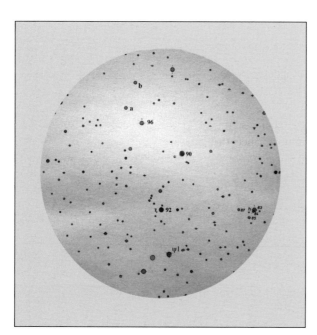

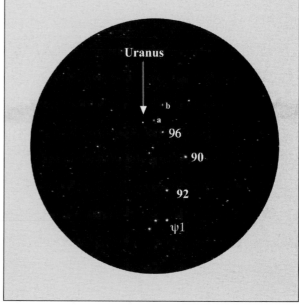

without optical aid. Uranus shines at magnitude 5.5 when brightest, and it has an average brightness of magnitude 5.7. In other words, Uranus can be seen by a person with average eyesight under a dark sky. But he or she has to know *exactly* where to look; even then it will most likely flit in and out of view as the eye tries to keep a focus on it, darting back and forth between averted and direct vision.

That also means that you must have a knowledge of the constellations and how to navigate the stars to find it. Usually, observers like to find the planet first in binoculars; they then put the binoculars away and test their naked-eye vision. Neptune, on the other hand, is 2.3 magnitudes fainter than Uranus when brightest. I do, however, believe the best and most advanced observers with excellent night vision can detect it with the unaided eye. Otherwise, binoculars are an absolute must for beginners to find it.

Close encounters (conjunctions)

One of the most enjoyable binocular sights is when two or more planets either pass near one another or the Moon. These alignments only fuel the imagination of the astrologers – especially when they form triangles or straight lines with one another in various astrological signs.

Perhaps the most famous conjunction of all (at least in the minds of some astronomers) is the one that may have occurred over Bethlehem during the time of Christ's birth. [Of course, explanations for the Christmas star run the gamut, from a divine sign to a nova, to a supernova (or hypernova), to a comet, to a stellar phrophecy, to a conjunction.] Adding to the confusion, there are also many conjunction theories. Johannes Kepler suggested it was a triple conjunction between Jupiter and Saturn in the year 7 BC, which he erroneously believed led to the creation of a nova that appeared over Bethlehem. Others have turned to the rare conjunction of June 17, 2 BC, between Jupiter (The King planet) and Regulus (the King Star) in Leo, the Lion. On that night, the two bright objects would have made an impressive naked-eye spectacle. Alas, this event occurred affter the death of Herod in 4 BC. More popular now is the conjunction between Jupiter and the Moon in 6 BC. This event culminated in the passage of the Moon in front of Jupiter (an occultation) on April 17. While not uncommon, this occultation occurred in the constellation of Aries, which, some believed, heralded the news that a divine king was to be born.

No matter, conjunctions are awe-inspiring events. During them, it's easy to see the varying speeds of these worlds as they march in step to the universal laws of gravity. When planets pass close to one another, you should also pay strict attention to their colors and see if you can't detect subtle differences. Monthly astronomy magazines, astronomy club newsletters, weather stations, and websites (like Spaceweather.com) keep you informed of upcoming planetary alignments. Amateur astronomy is a proactive hobby.

Ceres and the asteroids

The largest and first-known asteroid, Ceres, was promoted to the status of dwarf planet in 2006. Like our Solar System's eight planets, Ceres is a round body that orbits the Sun. It measures 933 km (580 miles) in diameter, making it about one-fifth the size of our Solar System's smallest planet, Mercury. Nevertheless, it may contain about one-third of the total mass of all known asteroids. Unlike the major planets, however, Ceres has not cleared the neighborhood around its orbit. Indeed, Ceres lies in the Main Asteroid Belt, which contains about 750,000 minor planets that measure 1 km (0.6 miles) or larger; about 100,000 of them are 10 km (6 miles) or larger. Thus Ceres joins the ranks of Pluto as being a dwarf planet: a round body that orbits the Sun, has not cleared the neighborhood of its orbit, and is not a satellite. Ceres, then, is no longer the largest asteroid but the smallest dwarf planet.

This is not the first time astronomers restructured the Solar System (nor will it be the last). Consider what has happened to Ceres over the years. When the Sicilian astronomer and monk Giuseppe Piazzi discovered this new world in 1801, the Solar System had only seven

planets. But since Ceres was at the right distance of the long-sought-after "phantom planet" – the one that predictions said should exist between Mars and Jupiter but could not be found – astronomers accepted Piazzi's discovery as a planet, which Piazzi named after the Roman goddess of the harvest and motherly love. Only a year after Ceres' discovery, Heinrich Olbers discovered another planet in the same region, a world he called Pallas.

That same year, however, William Herschel countered that Ceres and Pallas did not show disks like the other planets. Instead, he said, they "resemble small stars so much as hardly to be distinguished from them, even by very good telescopes." Thus, he create a new category of Solar System object to describe them: *asteroids* – after the Latin world *aster*, meaning star.

But in 1833, Herschel's son, John, reclassified Ceres, Pallas, and two other newly discovered asteroids (Juno in 1804, and Vesta in 1807) once again as planets. That brought to 11 the total number of major bodies in our Solar System. The discovery of Neptune in 1846 made it an even dozen. But by the middle of the nineteenth century, after astronomers began to discover more and more of these starlike worlds in the zone of the phantom planet, they decided to once again demote Ceres and its ilk to the class of asteroid. In the span of half a century, then, the number of planets in our Solar System expanded and contracted no less than five times.

Of course, Ceres is now no longer an asteroid but a dwarf planet. Hubble Space Telescope images and Keck images of Ceres' surface – which has an average

Ceres • January 24, 2004
HST ACS/HRC

Vesta • May 14, 2007
HST WFPC2

temperature of −73 °C (−100 °F) and absorbs 91 percent of the sunlight falling on it − shows a remarkably rich assortment of dark and bright features, which may be crater impacts, mineral deposits, or the effects of space weathering. It may also have a mantle of water-ice, and possibly a very weak atmosphere and frost.

The Main Asteroid Belt is a curious place. Just look at the average spacing between each successive planet in our Solar System and you'll notice a suspiciously wide gap between Mars and Jupiter − one large enough for a planet to occupy. Instead, what we find in that gap is a donut-shaped ring of "minor planets" orbiting the Sun at distances ranging from about 320 million to 495 million km (∼200 million to 300 million miles). They are not the floating wreckage of a shattered world, as some early theories argued. Instead, astronomers now believe them to be timeworn remnants of primordial material that Jupiter's gravity prevented from accreting into a planet when the Solar System was born 4.6 billion years ago. Had a planet formed in that gap, it might have been a body about 1,500 km (930 miles) in diameter − less than half the size of Earth's Moon.

You can see Ceres and several of the brighter asteroids in the main belt with handheld binoculars. To find them, you'll need to follow the same procedures outlined in the sections on Uranus and Neptune. Again, most astronomy magazines routinely publish detailed finder charts for the brightest asteroids when they are favorably placed in the sky. If you want periodically to check on the positions and brightness of your favorite bright asteroids, I suggest using the Minor Planet Center's Minor Planet and Comet Ephemeris Service (http://www.cfa.harvard.edu/iau/MPEph/MPEph.html). Just type in the name of the object you want to see in the appropriate box (and any date parameters), then click on "Get ephemerides," and voilà! . . . up comes the asteroid's position and magnitude (as well as other interesting orbital information). For precomputed ephemerides for bright minor planets, go to http://www.cfa.harvard.edu/iau/Ephemerides/ and click on the pertinent year listed under the heading, "Bright minor planets at opposition in."

The photo illustrations at right show the slow movement of a bright asteroid in a binocular field over the course of a day.

On page 145 you'll find a list of the brightest objects in the Main Asteroid Belt − those most commonly observed in handheld binoculars. They are arranged in order of mean magnitude at opposition. It is followed by a list of some of the more commonly observed asteroids for large mounted binoculars, ranging from those fainter than 9th magnitude to those brighter than 10th. The number preceding the object is the order in which the body was discovered; thus 1 Ceres, refers to the fact that Ceres was the first of these objects discovered, followed by 2 Vesta, and 3 Juno, etc.

Note that while Ceres is the largest body in the Main Asteroid Belt, it is not the brightest. Vesta, the third largest body, holds that distinction. Although Vesta is only about half as large as Ceres, its surface reflects 42 percent of the Sun's rays. Its orbit also brings it a tad closer to the Sun than Ceres; the combination of all these factors make Vesta more apparent to the visual observer on Earth.

Again, the lists give the *mean* magnitude of the objects at opposition. Many of the objects may become a full magnitude or more brighter than the listed value during their best oppositions. Vesta, for instance, can shine as brightly as a 5th-magnitude star, making it visible to the unaided eye! Interestingly, then, the most apparent object in the Main Asteroid Belt is not the dwarf planet that rules the kingdom, but the asteroid, Vesta, named for the Roman goddess of the hearth.

There's another class of asteroids that you'll want to keep an eye on: the Near-Earth Asteroids (NEAs). Not all asteroids orbit the Sun in the Main Asteroid Belt. Some get booted out of it through collisions with other asteroids, or by the gravitational influence of giant Jupiter. Every few months, astronomers catch one of these rogue objects

Brightest objects in the Main Asteroid Belt and common objects for handheld binoculars

Asteroid	Mean brightness (at opposition)	Brightness range
4 Vesta	6.0	5.3–6.5
1 Ceres	7.2	6.7–7.7
7 Iris	8.5	6.7–9.5
2 Pallas	8.6	6.7–9.7
15 Eunomia	8.9	7.9–9.9
8 Flora	9.0	8.0–9.8
3 Juno	9.1	7.5–10.2
6 Hebe	9.1	7.7–10.0
9 Metis	9.2	8.0–9.8
29 Amphitrite	9.2	8.7–9.6
18 Melpomene	9.4	7.7–10.4
20 Massalia	9.4	8.4–10.1
27 Eurtepe	9.7	8.4–10.6

Relatively bright objects in the Main Asteroid Belt and common objects for large mounted binoculars

Asteroid	Mean brightness (at opposition)
40 Harmonia	9.2
11 Parthenope	9.3
192 Nausikaa	9.3
10 Hygiea	9.5
39 Laetitia	9.5
16 Psyche	9.6
511 Davida	9.6
12 Victoria	9.7
13 Egeria	9.7
14 Irene	9.7

passing near or within the Moon's orbit shortly before or after it makes its closest approach to Earth. Some NEAs are potentially hazardous, because they might collide with our planet.

Most NEAs are on the faint end, being about 12th magnitude and fainter, and require a telescope to see well. But occasionally, about once or twice every decade, they get bright enough (around 9th magnitude) to be seen in binoculars. Large-binocular users should continually refer to websites like Spaceweather.com and Heavens Above for quick NEA alerts and finder charts.

Sometimes, we can use our binoculars to watch asteroids "colliding" with other celestial objects – namely stars. These are not real collisions but occultations. Each year dozens of asteroids pass in front of stars, the events of which are visible along long and narrow, and many times uncertain, paths across the globe. If you're lucky to be in the occultation path (and have good weather), you may see a naked-eye or binocular star disappear for a short period of time. Many times the asteroid is not even visible . . . but the effect of the passage is. These are very exciting and relatively rare events to catch.

Of course, catching an asteroid occultation requires that you know how to locate specific stars in the sky. One excellent website that will help you to get started in this exciting adventure is the International Occultation Timing Association's beginner's page: http://www.poyntsource. com/IOTAmanual/Preview.htm. As the site touts: "Never before has there been an opportunity to contribute to the body of scientific knowledge about the lunar limb profile and the size and shape of asteroids plus a host of other occultation phenomena."

I was fortunate to see one such event on the evening of March 23, 2003, from the the slopes of Mauna Loa volcano on the Big Island of Hawaii. That night the 10.7-magnitude asteroid (704) Interamnia passed in front of a 6.7 magnitude star. Although the asteroid was too faint to be seen in my 7 × 50 binoculars, I was able to see the star disappear and time it. But the star did not simply drop four magnitudes to the brightness of the asteroid (and thus out of my view) then reappear. Instead, after disappearing, it became faintly visible again before slowly swelling into view, dropping briefly in brightness once again, before regaining full brilliance – over the course of a minute. This was, in fact, one of the longest asteroid occultations on record!

Similar descriptions to mine were also made independently by some two dozen other observers in Hawaii and Japan. Consequently, Paul Maley of the Johnston Space Center, who spearheaded the observing campaign, summed up the observations: "One exciting discovery then is that we now suspect that the 6.7 magnitude target star HIP36189 is a likely close double star . . . the magnitude of the companion is +9.0 with a separation of only 0.016 arcseconds. The second discovery is that . . . the asteroid is a bit larger than previously thought." So real science can be accomplished with handheld binoculars! I strongly encourage you to participate in any decent event that crosses your area. It's truly fun and exciting!

One event that I missed but would have liked to have seen occurred on March 9, 2007 – when asteroid 3637 O'Meara crossed in front of a 10th-magnitude star, causing the star to drop 5.5 magnitudes (to the 15.5 magnitude of the asteroid) for a second over China and the Philippines.

If you're interested in asteroids, and just love to hunt them down (especially if you own large binoculars), you should consider joining the Astronomical League's Asteroid Observing Club: http://www.astroleague.org/al/obsclubs/asteroid/astrcobs.html. The Club issues awards to those who observe 25 asteroids! It may even lead you to purchase a telescope to hunt down ever-fainter objects.

CHAPTER 5

Comets: divine elegance

And David lifted up his eyes, and saw the angel of the LORD stand between the earth and the heaven, having a drawn sword in his hand stretched out over Jerusalem

I Chronicles, 21:16

Comets, those angels of the night, fly through the heavens on gossamer wings. No other nighttime Solar System object visible through binoculars compares in beauty to a bright comet with a long tail. But comets were not always perceived as beautiful. Ages ago, the sudden and surprising appearance of one had the power to corrupt man's thoughts, instill fear, and presage great calamities. To this day, comets continue to plague the minds of the uninformed, and they will probably continue to augur doom and gloom to the superstitious long into the future.

Before diving into the art of observing comets with binoculars, I thought you might first enjoy reading how some comets have (or may have) influenced humanity over the ages. For instance, it's arguable that the passage from the Old Testament cited above refers to a comet that appeared before King David, who, fearing retribution from the Lord for his great sins, interpreted it as a sign that God had sent an "angel unto Jerusalem to destroy it." As told in the first book of Chronicles, Jerusalem was spared when King David repented and the Lord forgave him.

Was this "drawn sword" really a comet? It's possible. In fact, it might have been a return of Comet 1P/Halley. As David Ritchie explains in his 1985 book, *Comets: The Swords of Heaven* (Plume, New American Library; New York), "Whether or not Halley's Comet was responsible for this incident from scripture, no one knows for sure; it was due for a return about the time of King David's repentance (somewhere around 1000 BC), but some other bright comet is also a possible explanation." The illustration below, from Johannes Hevelius' *Cometographia* (Danzig, 1668), shows the medieval perception of comets as a "sword of heaven."

Some comets in history

Great Caesar's ghost!

Shakespeare was no stranger to celestial signs and their influence on humanity. In one dramatic and often quoted passage in his play, *Julius Caesar*, we see Calphurnia, Caesar's wife, begging her husband not to leave the house because

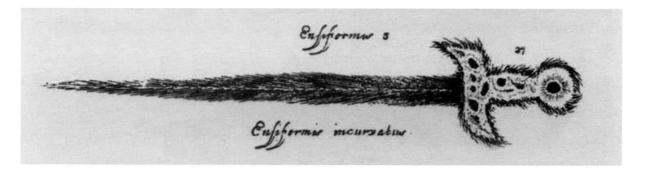

ill omens have appeared on Earth and in the sky. When Caesar refuses to listen, she pleads, "When beggars die there are no comets seen; The heavens themselves blaze forth the death of princes."

Although no comet burned in the sky when Brutus stabbed Caesar to death on March 15, 44 BC, one did blaze forth shortly afterwards, perhaps near the end of July. In his *Natural History*, Pliny the Elder (AD 23–79) tells us that the "late Emperor Augustus himself judged this comet highly propitious in that it appeared at the beginning of his principate, when . . . he was holding games in honour of Venus Genetrix, not long after his father Julius Caesar had died. He expressed his joy publicly in these words: 'On the very days of my games a comet was visible for seven days . . . It used to rise about an hour before dark and was bright and visible from all lands. The general populace believed that this signified that Caesar's soul was received among the spirits of the immortal gods . . .' This was what he said in public; inwardly he rejoiced because he interpreted that comet as having come into being for his benefit and as embracing his own birth."

Realizing the political advantage of the situation, Augustus shrewdly allowed Julius to become deified in the eyes of the awestruck Romans. Given the startling sepulchral nature of the spectacle, is it no wonder that the comet of 44 BC may also be responsible for the origins of the exclamation, "Great Caesar's ghost!"? That phrase was itself resurrected in the 1950s television series *Superman*; Perry White, the fictitious editor of the *Daily Planet* newspaper, constantly spouted that phrase whenever he was surprised or enlightened.

Medieval terrors

During the Middle Ages, comets had the power to fuel the imagination with wild visions of fantasy and horror. Indeed, in his 1907 book, *Popular Astronomy: A General Description of the Heavens* (Chatto & Windus; London), the French astronomy popularizer Camille Flammarion says that the "Middle Ages outdid, if this were possible, the foolish ideas of antiquity, and gave fantastic descriptions of certain comets which exceed anything which can be imagined."

During these dark times, comets were seen as swords of fire, bloody crosses, flaming daggers, spears, dragons, and mouths. Flammarion notes how "Paracelsus asserted that they were sent by the angels to warn us," and how the "sanguinary madman who was called Alphonsus VI, King of Portugal, hearing of the arrival of the comet of 1664, threw himself to the ground, loaded the comet with abuse, and threatened it with his pistol."

But few descriptions match the exorbitant imaginings of celebrated French surgeon and author Ambroise Paré (1510–1590), who, in his 1573 book *On Monsters and Marvels* (reprinted in 1982 by the University of Chicago Press), describes in gruesome detail the sight and effects of the comet of 1528:

This comet was so horrible and frightening, that it engendered such great terror in the common people, that some of them died of fear over it; the others fell ill. It seemed to be of an excessive length and so it was of the color of blood; at the summit of this [comet] one could see the shape of a curved arm holding a large sword in its hand, as if it would have liked to strike. At the end of the point there were three stars, but the one that was directly on the point was more bright and shiny than the others. On the two sides of the rays of this Comet there could be seen a great number of hatches, knives, swords, covered with blood, among which there was a great number of hideous *human faces* with their beards and hair bristling.

Of this passage, Flammarion concludes, "We see that the imagination has good eyes when it sets to work."

"Broom" stars

The illustration on page 105 is a broadside that depicts the comets of 1680, 1682 (P1/Halley), and 1683 in the sky over Augsburg, Germany, with the three horsemen of the Apocalypse (including a skeletal reaper). The eerie scene is framed in the face of a clock, the numbers of which are fashioned from weapons, bones, dismembered limbs, broomsticks, and a crucifix; the figures are separated by broken skulls. Note too how the tail of the comet of 1682 looks like an arm of the clock ready to strike the ominous hour of 13th.[1] Many of the weapons – the spears, swords, and maces – are especially symbolic of the fanciful shapes projected onto comets by the superstitious. Look, for instance, at the wavy blade of the sword representing the numeral I. The clockface itself is surrounded by four figures (the judgmental divine on top; the repentant below) accompanied by scripture.

What I find most interesting, however, are the straw broomsticks comprising part of the numeral VII on the clock's face. If the swords, spears, and other lance-like weapons of old can be inferred as comets, why not, then, the broomsticks, which, for centuries, have been a symbol of a witch's astral power?

The broom was used in some Wiccan rituals as a symbol of a sword. Comets have been seen as the swords of heaven. The parallelism between comets and witchs seems strong: an old European symbol of a comet is a bundle of straw; witches were once believed to fly through the air on their broomsticks, performing the work of the devil (bringing on plagues, storms, and many other natural disasters); comets with their broom-like tails were believed

---

[1] To the ancients, 12 was a perfect number (12 hours in a day, 12 months in a year, 12 zodiacal signs). By going beyond 12, to 13, one leaves the security of stability and enters the realm of instability and bad luck; Judas was the 13th person to attend the Last Supper; there are 13 witches in a coven; and 13 is the reversed date of Halloween – the witch's most sacred sabbat.

to bring about these same horrifying calamities, and were seen as works of the devil. Just look at the woodcut at the top of page 106, which shows the destructive influence of a fourth-century comet from Stanilaus Lubienietski's *Theatrum Cometicum* (Amsterdam, 1668); note how the tail is inverted like a broom, sweeping its horrible influence over the land.

In early modern Europe, the search for witches intensified during the Reformation (between the years 1520 and 1650). It is not surprising, then, that the fire-and-brimstone Puritan preachers who came to Salem, Massachusetts, after 1629 brought with them not only

their visions of a religious utopia for New England, but also the bloody superstitions of their motherland.

During the darkest days of young New England, Increase Mather, an erudite Puritan minister and the first president of Harvard College, used the shocking appearance of the Great Comet of 1680 to frighten the inhabitants of Boston and its surroundings into believing that the devil was alive in Massachusetts.

The comet, which was visible near the Sun in broad daylight, caused great alarm, especially in late December when it appeared after sunset with a long and "fiery tail" that stretched from the horizon to the zenith. In response to a growing panic, Mather published a sermon titled "Heaven's Alarm to the World," preached at the lecture of Boston in New England on January 20, 1680, in which he tells his flock, "There are also extraordinary stars sometimes appearing in the heavens... Such stars are called comets from the stream like long hair which useth to attend them." He warned that these "[f]earful sights" are verified in scripture, being "signs that the Lord is coming forth out of his holy habitation to punish the world for their iniquities." In preparation for God's action, Mather began to record "All, and Only Remarkable Providences," including "strange Apparitions, or what ever else shall happen that is Prodigious, such as 'Witchcraft' and 'Diabolical Possessions'."

Mather's obsession with signs was not overlooked by his son, the Rev. Cotton Mather, who, like his father, was scientifically astute. He knew of Newton's work on the Great Comet of 1680 and accepted that comets operated under natural law. But he also believed they possibly did

so through the agency of angels, or, even worse, *devils*. He believed that comets might be ministers of divine justice and could be habitations of animals in a state of punishment.[2]

Waiting for God's judgment, Cotton became feverishly obsessed with witchcraft, and, in 1688, condemned Ann "Goody" Glover, an "ignorant and a scandalous old Woman," for bewitching 13-year-old Martha Goodwin. When Glover refused to repent at Mather's bidding, she was hanged on November 16, 1688 − only two weeks after yet another comet appeared in the heavens.

After the hanging, Cotton took the afflicted Goodwin girl into his care. But when he saw her condition only worsen, he put pen to paper and wrote his magnus opus:

Memorable Providences, Relating to Witchcrafts and Possessions[3] − a work that many scholars say is the source for the witch hysteria of 1692.

Thus, the comets that appeared in the skies over Boston in the seventeenth century did not instigate the Salem witch hysteria (the causes are still being debated today), but their sudden and remarkable appearances were, without doubt, used by Puritan ministers in Boston and abroad as signs from God to justify their war against the devil.

Perhaps in a fitting close to this bleak chapter in New England history, in 1695, a great comet with a tail 30° long swept through the heavens like a witch on her broomstick. The apparition was strikingly visible on the night of All Hallow's Eve − our Halloween (an event not recognized or celebrated by the Puritans of the day).

One thing we do know for certain: records from the *ancient* far east (China, and maybe Korea and Japan) do describe comets as "broom stars." In China, their different forms were associated with different disasters. Interestingly, the Chinese also recognize a broom goddess, *Sao Ch'ing Niang*, who lives on the Broom Star and sweeps away rainy weather.

In War and Peace

While the comet of 1680 marked the culmination of comet superstition, comets continued to play with the minds of men. Consider C/1811 F1 (see page 107), the Great Comet of 1811, which developed a tail one hundred million miles long. Under its lustre Napoleon gathered his "grand armee," the greatest army assembled in Europe since Xerxes, and invaded Russia. The apparition was ominous and bore heavily on the minds of those going into battle. Indeed, after the invasion of Moscow, Napoleon ran short of supplies, and he and his men suffered great casualties owing to a harsh winter.

Leo Tolstoy (1828–1910) gave the comet recognition in his classic novel *War and Peace*; what he wrote is, in fact, my favorite literary passage concerning a comet, because it illustrates a turning point in our thinking about

---

[2] B-Horror-movie fans might be interested to know that these beliefs are mirrored in a 1961 classic, *The Brainiac*, about a witch who, while being burned at the stake in 1661, vows vengeance on his executioners. The witch returns 300 years later (in 1961) on a comet, whose orbital period is loosely modeled after that of the Great Comet of 1661, whose period is now on the order of 365 years.

[3] See http://www.law.umkc.edu/faculty/projects/ftrials/salem/asa_math.htm

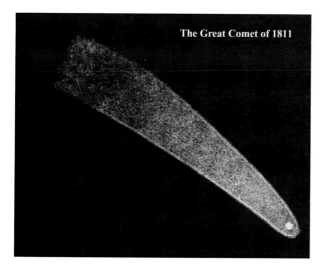

The Great Comet of 1811

them – showing superstition giving way to the beauty and harmony of science:

It was clear and frosty. Above the dirty ill-lit streets, above the black roofs, stretched the dark starry sky. Only as [Pierre] gazed up at the heavens did [he] cease to feel the humiliating pettiness of all earthly things compared with the heights to which his soul had just been raised. As he drove out on to Arbastsky square, his eyes were met with a vast expanse of starry black sky. Almost in the centre of this sky, above the Prichistensky boulevard, surrounded and convoyed on every side by stars but distinguished from them all by its nearness to the earth, its white light and long uplifted tail, shone the huge, brilliant comet of the year 1812[4] – the comet which was said to portend all manner of horrors and the end of the world. But that bright comet with its long luminous tail aroused no feeling of fear in Pierre's heart. On the contrary, with rapture and his eyes wet with tears, he contemplated the radiant star which, after travelling in its orbit with inconceivable velocity through infinite space, seemed suddenly – like the arrow piercing the earth – to remain fast in one chosen spot in the black firmament, vigorously tossing up its tail, shining and playing with its white light and the countless other scintillating stars. It seemed to Pierre that this comet spoke in full harmony with all that filled his own softened and uplifted soul, now blossoming into a new life.

The great white whale

Shortly after the Great Comet of 1811 faded from view, a new comet appeared in the night skies over the seafaring town of Nantucket, Massachusetts. In his enchanting 2000 book, *In the Heart of the Sea* (Viking Penguin; New York), author Nathaniel Philbrick explores how that

[4] C/1811 F1 peaked in brightness in the fall of 1811, but could be seen with the unaided eye for nine months.

whaling community was obsessed with omens. One sign was especially auspicious. In July, 1819, when the whaling ship *Essex* was being repaired and outfitted, a Quaker merchant, named Obed Macy, began following the comet from his house on Pleasant Street:

The comet (which appears every clear night) is thought to be very large from its uncommonly long tail, which extends upward in opposition to the sun in an almost perpendicular direction and heaves off to the eastward and nearly points for the North Star.

The New Bedford Mercury, the newspaper read by the Nantucket locals, offered that the comet was a sign of some impending "remarkable event." In the eyes of the Nantucket community, that event turned out to be the destruction of the *Essex* in August 1820 by an "enraged sperm whale." The famous marine disaster was, in fact, the event that inspired the climactic scene in Herman Melville's *Moby-Dick*.

But what of Melville's white whale itself? Melville's book is filled with celestial references, one of which I find most telling. In his description of the whiteness of Moby-Dick, Melville writes, "The flashing cascade of his mane, the curving comet of his tail...A most imperial and archangelical apparition..." Who could deny that a great comet, with its gently curving tail of dust and straight

spout of gas, looks like a great white whale breaching in a sea of stars.

World's end?

Time has failed to abolish age-old thinking about comets. In 1973, for instance, David Berg said that the Children of God cult announced the coming of what he called the Christmas Monster Comet, properly known as C/1973 E1 (Kohoutek). In the resurrected style of Increase Mather, Berg said the comet was a sign from God warning us to repent: "*There is going to be 40 days of warning, then Nineveh shall be destroyed! If the people repent, the Lord will relent or there'll be a descent!*" Nothing, of course, happened. Still, the hype generated by Berg had an influence on writers, such as Joseph F. Goodavage, who raised the question of the comet being a harbinger of doom. In his *Historical Dictionary of the 1970s* (Greenwood; Westport, Connecticut), he notes that *Time* magazine claimed the comet "might be a messenger – of light and knowledge for all mankind."

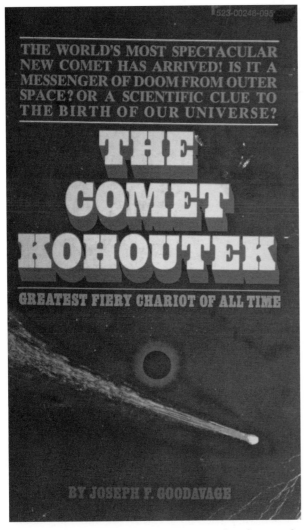

But Berg's antics were comical compared to the tragedy that occurred on March 27, 1997. That day, 39 members of the Heaven's Gate cult committed mass suicide in Rancho Santa Fe, California – so that their souls could hitch a ride on a spaceship believed to be hiding behind Comet C/1995 O1 (Hale–Bopp).

As long as there are comets in the heavens, there will be devils here on Earth.

The nature of comets and how to observe them

Alas, for all the fear they have caused, a comet is a minor Solar System body. At the heart of all that "fearful matter" is a tiny ice world measuring a mere 10 miles or less across. Most comets reside in what is known as the Oort Cloud. This vast sphere of icy bodies starts around 3,000 astronomical units (AU)[5] – about 75 times the distance of Neptune from the Sun – and extends out to 100,000 AU or 200,000 AU.

It's now generally believed that comets originated in the vicinity of the outer planets 4.6 billion years ago and were ejected by those planets outward. The NASA Deep Impact spacecraft image below shows the nucleus of Comet 9P/Tempel 1, which measures about 8 by 5 km. The data collected from that mission confirmed that the comet is "a relic from the formation of the Solar System."

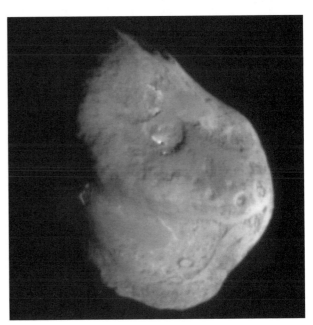

Because the Oort Cloud extends to the edge of the Sun's gravitational influence, its members are easily perturbed by passing stars and other forces, which can nudge them from their innocuous orbits in the cold depths of space and send them back into the inner Solar System along long and looping orbits. Indeed, the Oort Cloud is the primary source of *long-period comets* – those generally with orbital periods of hundreds, to thousands, to millions of years. *Short-period comets* (those with periods of hundreds of years or less), on the other hand, are

---

[5] An astronomical unit (AU) is a unit of length equal to the mean distance between the Earth and the Sun. It is roughly 150 million kilometers (93 million miles).

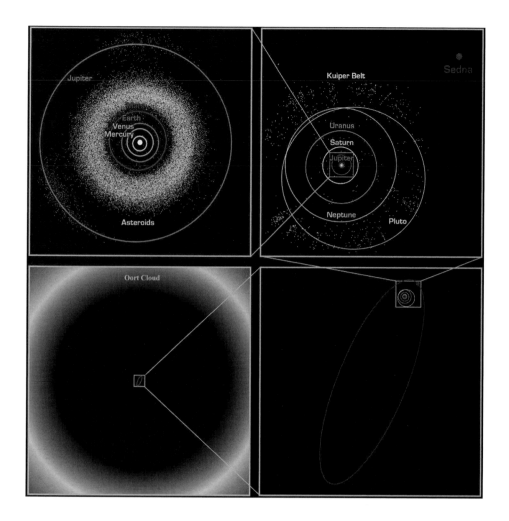

believed to originate from a roughly 1-billion-mile-wide, flattened section of the Oort Cloud that starts beyond the orbit of Neptune. Comet 1P/Halley, which orbits the Sun about once every 76 years, then, is a short-period comet.

Like planets, comets shine by reflected sunlight. When they are far away, they cannot be seen in binoculars. As they near the Sun, however, gases from the comet's nucleus boil off into space forming a *coma* – a glowing envelope of gas and dust that surrounds the nucleus out to 100,000 km (62,000 miles) or so in diameter. The effects of solar radiation pressure and solar wind can also push this material away from the Sun to form tails (these are described in more detail starting on page 112).

Keeping track of comets

Again, of all the wonders available to binocular users in the night sky, nothing compares to the beauty and grandeur of a naked-eye comet with a bright tail. Unfortunately, these spectacles are relatively rare, which is why the desire to see one can swell with passing time. According to the *International Comet Quarterly* (ICQ), on average, a comet brighter than fourth magnitude comes into view once every two years, and a comet brighter than magnitude 0 appears roughly once every 15 years.

Sometimes two bright comets can grace our skies in a span of a year, or 20 years can pass before we see such an apparition.

The point is, if you're interested in catching a comet with your binoculars, you have to keep abreast of what's happening in astronomy. Comets can brighten or appear suddenly with little or no warning. So it's important to be vigilant. The key place to begin looking for useful and accurate information regarding news, observations, orbital data, designations, and names – as well as good links regarding comets and related topics on the World Wide Web – is the ICQ's Comet Information website (http://www.cfa.harvard.edu/icq/icq.html). The ICQ also has a Headlines webpage for information on new comets: http://www.cfa.harvard.edu/iau/Headlines.html. Otherwise, the best way to keep informed is to subscribe to an amateur astronomy magazine, join an astronomy club (many clubs have event hotlines), search the World Wide Web for sites that feature daily updates on astronomical news (such as http://spaceweather.com), or frequent your local museum or planetarium.

Contrary to popular belief, comets do not streak across the sky. They rise and set daily like the Sun, Moon, and stars. Like planets, they move slowly and gracefully against the stars, changing their position little by little, night after night, until they disappear into the twilight or fade from view. Many bright comets remain visible for weeks, sometimes longer, giving us time to enjoy their beauty . . . weather permitting.

A bright comet with a prominent tail can be easy to spot under a dark sky away from city lights. All you need to know is when to look (for bright binocular comets it's usually either shortly after sunset or shortly before sunrise) and where to look. Glance skyward in the right direction at the right time, and you may see the celestial visitor as a ghostly white form hanging in the sky with a diaphanous tail pointing away from the Sun.

A binocular comet just below naked-eye visibility will require some careful searching. To find it, you'll need to know the comet's right ascension and declination, which you can plot on a detailed star chart. (Many astronomy-related websites post daily updates on bright comets and some plot the comet for you, which you can print out.) The image below shows a star-chart section from Wil Tirion's Sky Atlas 2000.0, on which I plotted the position of Comet C/1995 O1 (Hale–Bopp), then drew its

appearance as seen with the unaided eye. Note how the star chart has hours of right ascension (celestial longitude) at top and lines of declination (celestial latitude) at right.

Once outside, the star chart will help guide you to your target. Start by finding the brightest naked-eye star near the comet and place it in your binoculars. Once you are confident you have the right star, gently move your binoculars, field by field, in the direction of the plotted comet, until you intercept something fuzzy in form. But beware! Comets can pass near a bright deep-sky object, which may appear more conspicuous than the comet; it can be easy to mistake one of these wonders for your target. You could also use your detailed star chart to plan a "star hop" to the comet's location, which is a more precise way of hunting faint targets.

What to expect

What you see through your binoculars will vary from comet to comet and may vary dramatically from night to night. Comets, you will learn, can be very fickle guests: a faint comet may surge in brightness overnight, and a bright comet can fizzle in a matter of days. The view will also depend on the darkness of your observing site and the phase of the Moon. City lights, a neighbor's inconveniently positioned porch light, or a bright Moon, can greatly diminish the appearance of all but the brightest of comets.

Faint binocular comets may not have a conspicuous tail, appearing instead as a faint patch of diffuse light like that of an unresolved star cluster or a nebula. Unlike many comet photographs, which make the spectacle look like a burning fireball, the binocular comet is a much more delicate sight. A classic bright comet has three main structures: a starlike core (the *pseudonucleus*); a diffuse head (the *coma*); and a veil-like *tail*.

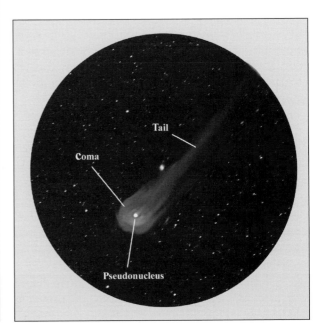

Note that I say *pseudonucleus*, not nucleus. By the time a comet is bright enough to be seen in binoculars, the true nucleus is masked from view by sunlight reflecting off the dust and vapors boiling off the comet's surface, which is why estimating the magnitude of the pseudonucleus over time may have special value. For instance, if the pseudonucleus suddenly surges in brightness, it may be due to an outburst of dust, caused perhaps as new vents on the comet's surface burst open. As a result, the starlike pseudonucleus may swell to a disk-like object as the erupted particles expand into space and reflect more sunlight. Sudden flare-ups may also be related to the fragmentation, or breakup, of the comet's fragile nucleus. Sometimes the pseudonucleus is too faint to be seen with binoculars – until a surge in brightness occurs, bringing it into view. Comet watching is full of such surprises.

Measuring the degree of condensation

Whether the pseudonucleus is visible or not, the comet's coma will have a measurable *degree of condensation* (DC). The measurement varies on a scale from 0 to 9, with 0 representing a totally diffuse coma with no detectable brightening from the coma's outer region to its center, and 9 representing a nearly stellar coma, where nearly all the brightness is contained at a central point or tiny disk. Binocular comets generally have DC values ranging from 0 to 7. The photo illustrations at right show comets with a DC of 1 (top), 5 (middle) and 7 (bottom).

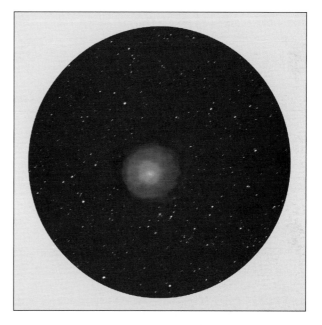

Estimating a comet's brightness

The comet's DC can affect the estimated brightness of the comet's total magnitude, which is much more difficult to do than making an estimate of a star. Unlike a star's brightness, which is a pinpoint of light, a comet's light is spread over a larger area of sky. Comparing the brightness of a comet with that of a star is a difficult process. The traditional method, the In-Out method, is to compare the size and brightness of the comet with a selection of similarly bright stars that have been racked out of focus until they appear the same size as the in-focus comet.

In another method, both the star and the diffuse object are simultaneously defocused and their brightnesses and sizes compared. The results of these methods can be quite accurate, but it has been my experience that they are inadequate for comets with strong central condensations surrounded by faint extended halos (comets with DC values of about 3). Simple in-and-out focusing methods lead one to underestimate the intensity of these awkwardly diffuse objects, because the contribution of light from the extended halo can be lost.

One solution to the problem is to use the Modified-Out method (conceived independently by comet observer Charles Morris and I). In this method, the observer

defocuses the comet until it appears as a uniform glow. The size and intensity of this view is memorized, then compared to views of comparison stars defocused to the same extent as the comet. The method is published in the *International Comet Quarterly*, volume 2, 1980.

By plotting a comet's magnitude and its DC over time, you can record how a comet's total magnitude is distributed across the coma, which may help comet researchers to better understand cometary dust and gas production. Comets with high DC values, for example, appear to produce a lot of dust or have very active nuclei.

Some beautiful features

As the activity on a comet's surface increases, especially around perihelion passage, the outflow of jetting particles may increase or become more regular. Because the effects of solar radiation pressure and solar wind is also stronger the closer a comet moves toward the Sun, the coma may develop a *parabolic hood* or *envelope* on its sunward side. Extremely bright or large comets may also show multiple envelopes inside the hood, which form as the comet's nucleus rotates. Very bright comets can also display thick *fountains* of light that fan outward from the nucleus toward the Sun. These may be accompanied by needle-like *jets* of material, which, like the fountains, can fan out before curving back into the comet's tail.

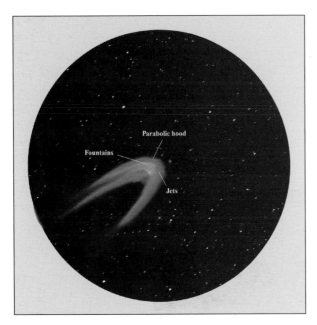

The parabolic hood also gradually blends into the comet's tail, which widens in the antisolar direction. Actually a comet has two types of tail. Most glorious is the gently curved *dust tail*, which shines by reflected sunlight; the tail curves because the dust particles released from the comet's nucleus follow the parent comet in its curved path around the Sun. The dust tail may be accompanied by a gas (*ion*) tail which appears as a nearly straight line extending away from the comet opposite the Sun. This ion tail consists of gases of water, carbon monoxide, and other simple atoms and molecules evaporated off the comet that have become electrically charged by interacting with sunlight. The charged material is then pushed along by the solar wind (a stream of charged particles that comes from the Sun and drags the solar magnetic field out into interplanetary space at a speed of about a million miles per hour).

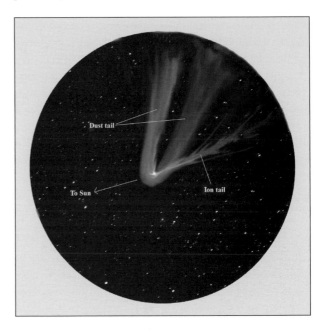

Experienced observers often see faint features in comets (like tails and jets) that less-experienced observers fail to notice. Seeing comet tails and other fine features is an acquired skill – a learned art that comes with practice and comet-observing experience, which is why you should try to observe as many comets as possible.

Most often, binocular observers (and certainly naked-eye observers) will see very little or no tail in a comet. Usually only the coma is visible, particularly if you're not observing from a very dark sky, not dark adapted, and not using averted vision to pick out usually dim tails and other features.

The rare bright naked-eye comets, and certainly great comets, are just awesome sights through binoculars. A bright gas tail may appear shredded into wispy *rays* of light, making it look like a ponytail of frayed hair pushed back by a strong wind. A high-contrast dust tail may curve like a scimitar. But again, even with bright comets, do not be surprised if both types of tails cannot be seen. Comets are as predictable as the weather. Sometimes only the bright dust tail will be obvious; sometimes the ion tail will dominate the view; some comets will toss out both types of tails with equal drama. These features can independently brighten and fade over time, so they require constant attention.

When a bright comet rounds the Sun, its tail may not only appear dramatically bright and curved but also striated. The origins of the striations, called *synchrones*, are not well understood, though the features may be the result of the Sun's magnetic field lines helping to organize the released dust particles as they interact with the solar wind.

A most spectacular display of synchrones occurred shortly after Comet C/2006 P1 (McNaught) reached perihelion on January 12, 2007. Around the time of perihelion, the comet flared to magnitude −6 and shone brilliantly in the daytime sky when it passed only 5° to 10° from the Sun. The comet's dashing tail spanned nearly a degree long and was visible to the unaided eye at high noon. Even the comet's parabolic hood stood out in broad daylight to the unaided eye. The spectacle was most fantastic in 10 × 50 binoculars, which showed the object against a crisp blue sky as a white star with a tail nearly two apparent Moon diameters in length, as seen in the photo below.

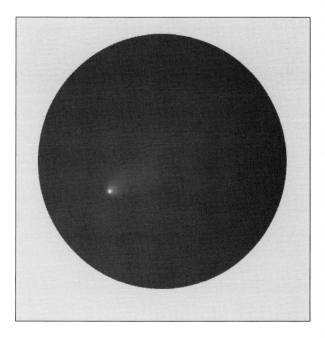

After the comet slipped below the horizon for Northern Hemisphere observers, it became a stunning visual apparition in the Southern Hemisphere. When my friend Ian Cooper of New Zealand first spotted the tail in the twilight, he said, "the tears started to flow."

The comet's tail was so long and curved that the striations were visible for weeks from the Northern Hemisphere. I was first made aware of the synchrones on January 18, after a friend told me to look at the Spaceweather.com website, at a photograph taken by Colorado amateur Daniel Laszlo. The photograph immediately reminded me of the famous woodcut of C/1743 XI, more commonly known as Cheseaux's Comet of 1744 (below) – something I've wanted to see all my life but thought I never would.

Inspired by Laszlo's photograph, I drove to the 7,000-foot-level of Hawaii's Mauna Loa volcano after twilight and was not prepared for what I saw. No one had told me much about the visual brightness of the synchrones, nor of their apparent size. I was expecting to use keen averted vision and to see something extremely faint and structurally weak, especially since it was going to be projected against the zodiacal light.

The drive up the mountain was a nail-bitting experience, since the clouds hugged the car as it climbed, and they did not disperse until I got within a few hundred feet of the observing site. I immediately got out of the car and nearly feel to my knees. The synchrones were fantastically bright and gorgeously large (see the photo on page 114).

The synchrones were clearly an order of magnitude *brighter* than the zodiacal light, which is at its brightest and most vertical this time of year from the subtropics. They appeared as a half-dozen or more phantom brush strokes against a rich canvas of smoking starlight. The longest stretched some 25° to 30°, and there must have been even more below the shoulder of the mountain. I cannot begin to impress upon you the glory of the sight, which is unlike anything I have ever seen. I have seen the synchrones in the tail of comet C/1975 V1 (West), but to see synchrones from a comet below the horizon sticking

up like feathery moonbeams was quite an unexpected and pleasant surprise.

When Barbara Wilson of Houston, Texas, combined a photograph taken by the comet's discoverer (Robert McNaught in Australia) and one of mine taken from Mauna Loa, she noticed something familiar about the comet's appearance:

> When I combined the two photos, my husband, Buster, and I immediately saw the same thing: an Indian chief's feather headdress adorned with eagle feathers. Then the thought came that maybe the eagle feather design may have come from something buried long in their history, perhaps a great comet.

Once a comet rounds the Sun, the nucleus heads into the tail. If the Earth happens to cross the plane of the comet's orbit shortly after the comet rounds the Sun, we may see an edge-on view of dust particles in the sunward direction, making the comet appear to have a sharp lance of light pointing toward the Sun from the comet's head. Since this ray of dust extends in the opposite direction from the comet's tail, it is called an *anti-tail*.

Some wonderful comets

Here is a short list of some beautiful naked-eye and binocular comets that have appeared over the last century. Of those that I've observed in the last three decades and more, I cannot recall one being a disappointing sight.

Comet C/1910 A1 (Great January Comet)

Workers at the Transvaal Premier Diamond Mine in South Africa first discovered this comet on January 13, 1910, with their unaided eyes. The comet was in the dawn sky and appeared like a star with a tail. By January 17, Northern Hemisphere observers could see the comet with their naked eyes less than 5° from the Sun's limb in broad daylight, appearing as a "snowy white object" with a tail 1° long! The comet peaked at magnitude −5, making it only slightly brighter than the planet Venus at maximum brightness. But I suspect that the comet had to be much brighter, perhaps as bright as magnitude −6 or −7, because Venus cannot be seen that close to the Sun without optical aid.

When the comet reached the evening twilight, its head looked about as bright as Venus, and it sported a 10° tail. Writing in a 1912 Lowell Observatory *Bulletin*, Carl Otto Lampland said that, during the last days of January, the comet was a "splendid object" to the naked eye, "and the grandeur and beauty of the spectacle were increased by the nearness of Venus and the bright cone of the zodiacal light."

Comet 1P/1909 R1 (Halley)

Comet Halley made a fantastic appearance in 1910, becoming visible to the unaided eye as an approximate 2nd-magnitude "star" with a short tail. The great observer Edward Emerson Barnard followed the comet with his unaided eye from April 29 to June 11. When brightest, the comet appeared in the morning sky as a brilliant 0-magnitude object with a tail that stretched up to 120° across the sky. Recollecting his impressions in a 1914 *Astrophysical Journal* article, Barnard said that "to the human eye it left a lasting impression which, added to its long life-history of more than two thousand years, made it, at its return in 1910, perhaps the most interesting comet of history."

Curiously, Barnard noted that "the newspapers and the general public were disappointed in the comet," noting that the probable causes were the "smoke and electric lights of today," which he said "robbed the comet of its glory," combined with the fact that sensational newspaper coverage had "excited the public pulse to a high pitch . . . beyond all reason." Barnard lamented, "It would have been a gratification to know that everyone who saw this wonderful object, saw it with the same spirit of elation and wonder − one would almost say veneration − with which the average astronomer greeted it. This was, at least the feeling of the present writer when he looked at this beautiful and mysterious object stretching its wonderful stream of light across the sky."

When Comet 1P/Halley made its second return in the twentieth century, I first recovered it visually in January 1985, using a 24-inch reflector atop Mauna Kea, Hawaii. The comet's magnitude was only 19.6. By year's end, the comet was just becoming visible to the unaided eye and was a spectacular object in 10 × 50 binoculars under dark skies. The head, though small and starlike, had a dense punch to it, perhaps a direct result of an apparent outburst of dust from the nucleus that occurred in mid December, 1985. Around that time, the comet's ion tail had a *disconnection event*, which by month's end was keenly apparent to the binocular user.

Disconnection events are one of the most spectacular phenomena observed in comets, and are generally accompanied by visible knots, kinks, and other curious structures. On December 28, for instance, a new sharply defined ion tail could be seen emanating from the comet's dense head. A sweep of the binoculars up the comet's 5°-long tail revealed that the new ion tail was followed by a dense knot of material about a third of the way up the tail followed by gap between it and the disconnected tail. All of this was tightly wrapped in a thin, yet sweeping dust tail. A bright Moon fought the comet at it brightest; it then sailed south and became a Southern Hemisphere object, where it remained a somewhat underwhelming sight, showing a broad fan of dust and a stumpy ion tail. Still, the comet was a fine binocular object, looking somewhat like a badmitten birdie.

Comet C/1956 R1 (Arend–Roland) and Comet C/1957 P1 (Mrkos)

These two comets were the naked-eye marvels of the 1950s. Belgian astronomers Silvio Arend and Georges Roland discovered C/1956 R1 on photographic plates in September 1956. At the time, it shined at 10th magnitude. But by mid April 1957, the comet had become a 0-magnitude beacon in the fading dusk. Aside from its stunning 30° tail, the comet also had an equally bright (and half-as-long), sunward-pointing tail, or *anti-tail*. Just as memories of C/1956 R1 were fading, Comet C/1957 P1 made a surprise appearance in the dawn. Czech astronomer Antonin Mrkos discovered this comet with his unaided eyes when only its 15°-long tail was peeking above the horizon; shortly after, the comet's head rose just prior to sunrise! The Comet C/1957 P1 (Mrkos) soon became a brilliant two-tailed spectacle in both the morning and evening skies. Although the comet's brightness had dropped to magnitude 3 by mid August and then continued to fade, the comet remained faintly visible to the unaided eye until mid September.

Comet C/1965 S1 (Ikeya–Seki)

This comet is one of my childhood memories and mysteries. At the impressionable age of eight – and only a few weeks before my ninth birthday – I was looking out an open window just before dawn, when I saw an odd searchlight beam in the southeastern sky. It was in the direction of the Star Market, a local supermarket, but I couldn't understand why anyone would be using a spotlight at this time of day. Halloween was upon us, I reasoned. Still, everything about that morning had an aura of mystery: the blood red glow of an approaching dawn, the bite of the autumn air, the dying light of stars . . . all punctuated by the hypnotic vision of a strong vertical slash of light.

Not until 1971, when I was sifting through the logbooks of the Harvard College Observatory, did I find a solution to that childhood mystery. The spotlight had to have been the long and luminous tail of Comet C/1965 S1 (Ikeya–Seki).

Two Japanese amateur astronomers, Kaoru Ikeya and Tsutomu Seki, independently discovered the comet in their telescopic sweeps of the heavens on September 18, 1965. It went on to become the most brilliant comet of the twentieth century. On October 21, when the comet was closet to the Sun, it shone at a peak magnitude that ranged from −10 to −15! Expert observers spotted it only 2° from the Sun's limb with their unaided eyes, simply by blocking the Sun with the palms of their hands. This awesome sungrazer then rounded the Sun and entered the early morning sky in the southeast. Around the time when I was looking out that window, the comet's tail (a thin saber of light) was near its maximum length of 60°!

Comet C/1969 Y1 (Bennett)

Discovered by the South African comet hunter John Bennett on December 28, 1969, Comet C/1969 Y1 began its journey as tailless 8.5-magnitude binocular object. But it steadily brightened and became a naked-eye wonder by late March 1970, when it shone at 0 magnitude and had a striking and highly structured 10° tail that blossomed to twice that length in the following month. By month's end, the comet had faded to about 4th magnitude but it kept its lengthy tail; around this time, the comet also became circumpolar and remained visible all night!

Comet C/1973 E1 (Kohoutek)

Billed as "The Comet of the Century" by the press, Comet C/1973 E1 was a major public disappointment. That said, it was actually quite a surprising sight in binoculars around the time of its perihelion, which occurred on December 26, 1973. It was my first binocular comet. The comet reached perihelion the next day, and I saw its subtle glow above the treetops from our second-floor apartment in Cambridge, Massachusetts. The comet was shining around 4th magnitude and had a very broad and ghostly tail, appearing very much like a sixteenth-century engraving of a comet.

Comet C/1975 V1 (West)

This comet will always live long in my memory, because it was my first great comet (that I was conscious of). Richard West discovered the dim 15th-magnitude object on August 10, 1975, while surveying photographic plates taken with the 100-cm Schmidt camera at the European Southern Observatory in La Silla, Chile. The comet brightened significantly as it approached both the Earth and Sun, and on February 25, 1976, I detected it through Harvard

College Observatory's 9-inch refractor a few minutes after local noon, making it the first visual and ground-based comet since C/1965 S1 (Ikeya–Seki) was observed in broad daylight. My friend and observing partner, Peter Collins, saw it immediately thereafter in the telescope's 3-inch finder; as the Sun neared the horizon, I spied the comet in 7 × 35 binoculars, appearing as a dashing white form with a degree-long tail directed straight up from the Sun like a woman's white bridal train. The comet's head was quite condensed and planetary in appearance, shining then at magnitude −2.

Despite that magnificent show, nothing prepared me for the comet's appearance on March 3, when I went out before sunrise and saw from my backyard the comet dangling above the eastern horizon like a bloody scimitar. In one shocking moment I understood the long history of comets as portents of evil and the death of kings. For the next week, the comet's dust tail grew in prominence (up to 26° as seen from Cambridge) and broadened into a magnificent fan, laced with multiple synchrones. The gas tail shot straight up as a pale blue streak, and through the 9-inch, I watched the comet's nucleus split into four fragments, two of which could be seen in binoculars until late June. The one thing I've learned since observing this magnificent spectacle is that great comets hold great surprises.

Comet C/1983 H1 (IRAS–Araki–Alcock)

This comet was one of the strangest naked-eye and binocular comets I've seen. The comet was first imaged on April 25, 1983, by the Infrared Astronomical Satellite (IRAS), then discovered visually on May 3 by two amateur astronomers: Genichi Araki in Japan, and George Alcock in England; Alcock, 70 years old at the time, discovered the 6th-magnitude object with a pair of 15 × 80 binoculars while kneeling on the floor of his house and scanning the skies in Draco through a closed window.

On May 11, the comet passed closer to Earth (2.9 million miles) than any comet since Lexell's in 1770, and that historic event sent me (and many other observers around the globe) scrambling for a view under a dark sky. The 1.7-magnitude coma spanned some 7 Moon diameters (if placed halfway between the pointer stars of the Big Dipper, the comet's coma would have extended 1° beyond either star). Because it was so large and diffuse, making a magnitude estimate required defocussing stars with unaided eyes! In binoculars, the comet's coma took up a substantial portion of the field of view and appeared as a bulbous onion head with multiple hoods, three stringy sunward jets, and a tailward spine that looked something like a rat's tail; it was an intrinsically faint comet, however, so no dramatic tail was visible. The comet moved so fast (relatively speaking) that substantial motion could be detected in binoculars in a matter of minutes.

Comet C/1996 B2 (Hyakutake)

This comet was arguably the most awe-inspiring sight in the deep night sky in recent memory. I say "deep night sky" because when the comet was at its best in March 1996, it sailed high in the midnight sky sporting a tail that stretched for more than 100° in length. Japanese comet hunter Yuji Hyakutake discovered it on the morning of January 30, 1996, using a pair of 6-inch binoculars – the kind you find on battleships. Just a dim 9th-magnitude glow at first, the comet gradually grew in brightness, and by mid March was a fine binocular object with a dashing ion tail. But as the comet passed to within only 9.3 million miles of the Earth on March 25, no one expected to see such beauty and grace. Shining at magnitude −0.8, the comet's head (which had an eerie, green sheen to it as seen through binoculars) swelled to some four Moon diameters, while the ion tail stretched more than half way across the sky, appearing intensely frayed through 7 × 50 binoculars.

Most great comets appear best only when seen near the Sun shortly before or after twilight, but Comet Hyakutake (C/1996 B2) was best seen while high overhead in the middle of the night! When my wife saw it from Volcano, Hawaii, she thought it was a moonbow. Another friend told me that when he stepped outside and glanced up, he fell to his knees in awe of the spectacle. Through 7 × 50 binoculars, the comet's pseudonucleus, jets, fountains, and ion tail (complete with hairy split ends) could be monitored for about a month in crystal clarity. When closest, the comet was moving about 1° per hour, so binocular users could also watch the specter fly against the stars over the course of hours.

In yet another rare visual display, on March 25, the comet's ion tail broke off from the comet's head and reformed a few hours later – a disconnection event.

Comet C/1995 O1 (Hale–Bopp)

This comet remained bright and easily visible for so long (18 months) that it became almost a fixture in the visual nighttime landscape. When Alan Hale and Thomas Bopp discovered the comet telescopically on July 23, 1985, it was a faint 11th-magnitude object more than 600 million miles from the Sun! For an object so small to appear so bright at such a distance presaged greatness for this comet. Indeed, from the time the comet became a binocular object in March 1996 to the time it faded from binocular view in early 1998, observers all around the world had ample opportunity to study the comet's many magnificent details – which were apparent even under light-polluted skies. These included the dust and ion tail, jets, fountains, parabolic hood and envelopes, the pseudonucleus, and colorful yellow coma. In fact, the comet's head was so dense and the pseudonucleus so bright, that when it shone around 2nd magnitude in February, 1997, the

comet's pseudonucleus, parabolic hood, and some jets could still be seen in a 4-inch telescope just 20 minutes before sunrise! In binoculars, the comet was a stunning textbook example, sporting a dense and gently curving yellow dust tail that spanned some 30° of sky, accompanied by a straight blue gas tail that shot up from the horizon like water from a spouting whale.

Comet 73P/Schwassmann–Wachmann

Discovered by German astronomers Arnold Schwassmann and Arno Wachmann in May 1930, Comet 73P/Schwassmann–Wachmann is a member of Jupiter's family of comets.[6] Its tiny nucleus, which measures no

more than about two-tenths of a mile in diameter, orbits the Sun every 5.3 years and is usually a decent telescopic object. But in the fall of 1995 the comet suddenly flared in brightness, bringing it to the threshold of naked-eye visibility and making it a fine comet in binoculars. The reason for the flare-up was a fracturing of the nucleus into several pieces. The image above shows a Hubble Space Telescope view of the amazing breakup.

Then, in April and May, 2006, the comet put on a spectacular binocular show with two of its many fragments being widely visible in the nearly same binocular field as they sailed through Corona Borealis, then by the Keystone of Hercules. In 7 × 50 binoculars the two fragments (both shining at around 7th magnitude) and their tails looked like sisters: one slender, one plump; one wearing a long dress, the other a wide skirt. I watched these sisters perform in late April with some 500 participants at the Texas Star Party in southwest Texas. Even the dim binocular comets can have special charm. I only hope that you get to experience many of your own.

[6] A Jupiter-family comet is one with an orbital period of less than 20 years, a semi-major axis shorter or roughly the same as that of Jupiter's orbit, and an orbit of low to moderate inclination, typically less than 40°.

CHAPTER 6

Meteors: when the heavens weep

Now slides the silent meteor on, and leaves
A shivering furrow, as thy thoughts in me.
                    Alfred Lord Tennyson, *The Princess* (1847)

Meteors, meteoroids, and meteorites

Meteors have been known throughout time as shooting, or falling, stars. But they are not stars at all. In my youth, before I knew any better, I truly thought that meteors were dying suns . . . that somehow, just before a star exhausted its fuel, it perished in a blaze of glory, speeding through space like a flaming balloon whose air had been released suddenly through its nozzle.

The truth is not so dramatic. *Meteors*, those swift streaks of light in the night sky, are the glowing paths of tiny solid particles (*meteoroids*) that plunge into Earth's upper atmosphere from space. Seeing a meteor is an infinitesimal example of what happens when worlds collide.

The particles responsible for the vast majority of meteors are sand-sized bits of comet crumble – loose lattices of dusty material (with textures similar to that of instant coffee). Many of the larger particles (>1 centimeter across) are probably related to asteroids, which have more solid structures, allowing them to burn longer in our atmosphere; a tiny fraction of these may have been blasted from the Moon or Mars during collisions with large meteoroids or asteroids in the dim and distant past.

Meteoroids are so prevalent (countless billions) that as the Earth orbits the Sun, we collide with untold millions of them each day. Collectively, that's enough solid matter to weigh some 55,500 kilograms ($\sim$120,000 pounds) – roughly equal to that of ten adult Asiatic elephants.

When a meteoroid enters our atmosphere, it does so with tremendous speed: up to 72 km (44 miles) per second for a head-on collision.[1] As it plunges, the particle rubs against air particles, creating friction, which causes the object, and the air around it, to glow. (Rub your hands

---

[1] As it orbits the Sun, the Earth travels at about 29 km (18 miles) per second. The fastest meteoroids travel in their orbits at speeds of 42 km (26 miles) per second. Thus, when a meteoroid meets the Earth head-on, the speed of the impact is the sum of their velocities: 29 + 42 = 71 km per second; Earth's gravity adds another km per second.

fast and hard against your legs and you'll feel the heat, though that's nothing compared to what the meteoroid experiences.)

Meteors become visible at altitudes ranging from 120 to 65 km (75 to 40 miles) – where the air is dense enough to cause resistance; the particle typically vaporizes at altitudes of 95 to 50 km (60 to 30 miles). An average meteor (one most likely to be recognized by the public) shines around 2nd magnitude, about as bright as a star in the Big Dipper, or Plough. The light show is over in a matter of seconds, or a fraction thereof.

Actually, the streak we see is an illusion. It's a combination of the object's great speed and the phenomenon known as the *persistence of vision*. Our eye/brain system cannot detect rapid changes in brightness. When the retina captures an image, it retains it for only one-tenth of a second before processing the next one. If separate images are flashed before the eye in rapid succession (more than about 10 per second), the brain thinks it's seeing a continuous scene. This is how movies work. Meteor streaks, then, are the sky's most celebrated cinematic sensations!

Fireballs

Not all meteoroids are small. They can range in size from anything larger than a clump of molecules to about the size of a small asteroid (a few tens of kilometers in diameter). A fragment about the size of a fist can produce a light show about as bright as the full Moon. Any meteor brighter than Venus (magnitude −5 or brighter) is called a *fireball*. A person on the street might be lucky to catch one fireball in a year or two . . . if they're in the habit of looking up! But meteor observers, especially those who

spend time watching meteor showers, can see many, many more. Regardless, the sight of one is not often forgotten.

Under a dark, moonless sky, a fireball can bathe the landscape in an eerie light bright enough to cast shadows. Because the falling fragment is relatively large and penetrating the atmosphere at great speed, fireballs usually remain visible for several seconds before dissipating. This gives the prepared binocular user time to follow it to its heavenly grave. As the objects blaze across the sky, it's common to see them suddenly and unexpectedly flare in brightness, perhaps several times before winking out. Sometimes, as they descend, small "molten blobs" bleed from their bulbous heads. Others may sparkle, spit, or even burst in mid-flight.

Meteor trains

The very brightest and very fastest meteors leave in their wakes a *meteor train* – a cylindrical trail of ionized atoms that remains in the meteor's path *after* the meteor has passed. Some forbidden lines in oxygen tend to glow longest and cause the train to glow green. Many trains last less than a few seconds, and these will be a challenge to see through your binoculars before they fade.

A train that lasts longer than 10 seconds is called a *persistent train*. A persistent train gradually becomes more intense as the green meteor train fades; it also turns orange as ozone diffuses into the wake and the reaction creates iron oxide (FeO) in an excited state. Some will last many minutes, while those from the brightest events may last hours; some of these will penetrate to the lower regions of the atmosphere, allowing them to be seen in the sunlight, appearing as twisted trails of smoke and dust.

Persistent trains are a phenomenon not to be missed by the binocular observer, particularly because they constantly morph before your eyes, contorting into all manner of shapes by the hurricane-speed winds in the upper atmosphere.

I witnessed one such event on the morning of November 17, 2006. The drama began as a magnificent fireball descended from high overhead and exploded in complete silence about 20° above the eastern horizon. I shared the wonder of the visual spectacle in my Secret Sky column in *Astronomy* magazine (November 2007):

> After the flash, I saw the spirit of the fireball dancing on its grave. It appeared as a luminous cloud left behind in the meteor's wake. Straight and sharp, the phantom train lanced the stars above the tree line, where it remained eerily suspended like an icicle waiting to fall. Suddenly, the train buckled like a boxer who received a sucker punch to the stomach. Over the next few minutes, the train faded to black. The "house lights" on this naked-eye performance went down and the curtain was drawn.

Although the train vanished to the naked-eye, it remained clearly visible in 10 × 50 binoculars for about a half

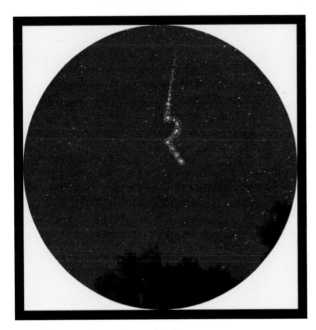

hour. The photo illustrations here show how the train changed over time: first fracturing into a beaded ring, then morphing into a diffuse butterfly, then mimicking a binocular comet with two tails.

These shapes appear to be common in long-lived events. For instance, in a 1907 issue of the *Astrophysical Journal*, Charles Christopher Trowbridge says that on November 14, 1868, an observer in Madrid, Spain, described a large fireball that "left a train visible ten minutes which expanded 6°–8° in width, then faded out in the center so as to form a ring." He also notes that "Professor E. E. Barnard observed a train in November 1901, which gradually expanded so as to appear 'like a comet with a double tail.'"

Indeed, as my photo illustrations show below shows, toward the end of the observation the train appeared as two linear "comet tails" running parallel to the horizon, both of which ultimately faded.

It's not uncommon for a single persistent train to evolve into two parallel ones, especially if it is observed through binoculars. The phenomenon has been noted for well over a century. As early as 1868, Hubert Anson Newton of Yale University opined that the double train is "due to the actual duality of the meteor itself." Trowbridge countered that more likely "the train gradually becomes a tube of luminous matter, which, viewed from the side, appears like a double line of light. The effect may be caused either by the dying-out of the luminosity along the axis of the train, or a greater luminosity at the border."

In his 1907 *Astrophysical Journal* article (volume 26, page 95), Trowbridge provides several illustrations of double trains observed during the great Leonid meteor shower in November, 1868 (the left illustration on page 118), a sample of which is shown below. They were originally drawn by Winthrop Sargent Gilman, Jr., of the Palisades Observatory in New York, where the trains were observed with a small telescope. The drawing at left shows a naked-eye view of a persistent train that appeared as "a double line of bluish-green luminous matter" through a telescope (right), leading the author to conclude: "It is evident from the foregoing that many meteor trains which would appear to the naked eye like a single bar of luminous matter might exhibit a dual appearance if observed through a small telescope." This point emphasizes the importance of today's observer viewing meteor trains with optical aid, including binoculars!

Persistent and double meteor trains still remain somewhat of a mystery. In a 2003 *American Geophysical Union* abstract, Michael C. Kelley and his colleagues proposed that the phenomenon is related to the sedimentation of dust particles produced in the train – an idea supported, in part, by rocket-based observations. In a 2005 *Journal of Geophysical Research* abstract, however, John Zinn and Jack Drummond of the Los Alamos National Laboratory, tell how they successfully modeled the meteor train as a hot cylinder that first expands radially and then rises because of buoyancy. As it rises, the cylinder evolves into a pair of counterrotating linear vortices: the double train. Detailed observations of persistent meteor trains are needed, and binocular observers making careful records could help to refine these theories.

Bolides

Very bright fireballs that not only explode in the air, but also produce a sonic boom, are called *bolides*. They're one of the most dramatic and awe-inspiring events in the nighttime sky. I was fortunate to have witnessed one such event on the morning of January 14, 1999. Around 3:45 am I was in my yard in Volcano, Hawaii, loading my observing gear into a suburban, when suddenly I saw a bright orb low in the west through the rear windshield. At first I thought that a neighbor had installed a new floodlight, but then the landscape around me began to brighten until it looked much brighter than that illuminated by the full Moon. Instinctively, I looked overhead and saw a flaming fireball – silver with a blue fringe – swelling enormously. Suddenly the fireball flashed so violently (like a welder's torch turning on) that I quickly looked away. When I returned my gaze, I saw the fireball tearing up into colorful fragments. The pieces flew off in all directions with smoky trails. The explosion reached at least magnitude −20 and came at approximately 3:47 am. The bolide's head continued to fall eastward until it reached an altitude of about 50°, when its light was "snuffed out" near Alpha (α) Bootis (Arcturus).

About 10 seconds later, a loud explosion caused me to flinch. The noise was not like that of a volcanic explosion (which I have heard many times) or of thunder, but more

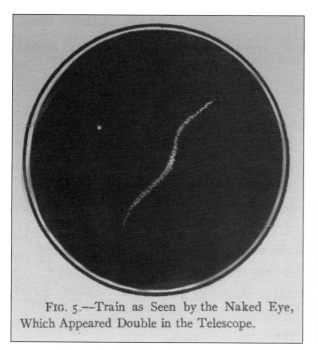

FIG. 5.—Train as Seen by the Naked Eye, Which Appeared Double in the Telescope.

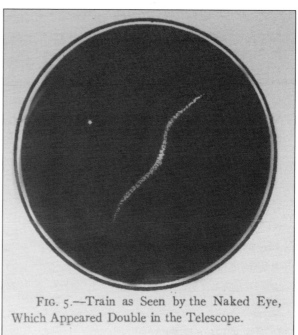

FIG. 5.—Train as Seen by the Naked Eye, Which Appeared Double in the Telescope.

muffled, like that of a depth charge detonating beneath the sea. That initial explosion was followed by 30 seconds (I counted) of rumbling, like rolling thunder.

Many Big Island residents were awakened by the brilliant flash and *boom*, which rattled windows in some homes. The shock of the explosion also triggered all the seismographs on the island.

This bolide likely generated shards large enough to survive the re-entry burn and rain down on Earth's surface as *meteorites*. Indeed, every few years, researchers recover pieces of rock from outer space shortly after they have fallen to Earth after the grand sky show. In this case, however, the meteorites, which are always covered with a black (or rusty brown) fusion crust, would have fallen either onto the active lava fields of Kilauea Volcano, or into the surrounding sea. Finding a fragment from this fall would be extremely difficult at best.

While many small- to medium-sized meteorites fall to Earth each year (three-fourths of which end up in Earth's oceans), it's estimated that a meteorite with a mass of about 1,100 tons will strike the Earth every 20 to 30 years — large enough to leave a crater the size of a football field . . . if it reaches solid ground. One such impressive feature is Meteor Crater near Winslow, Arizona (see photo above); it is the first confirmed, most popular, and best preserved impact feature on Earth. It was created some 50,000 years ago, when a nickel–iron meteorite — measuring about 24.5 meters (80 feet) across and weighing roughly 300,000 tons — slammed into Earth at a speed of 65,000 kilometers (40,000 miles) per hour. The impact left behind a hole measuring 1.2 km (0.75 miles) across and nearly 200 meters (600 feet) deep.

No solid evidence exists of anyone being killed by a meteorite. Allegedly one struck and killed a monk in Cremona, Italy, in 1511; suspiciously, however, another monk was supposedly killed in the same fashion in Milan 139 years later.

Some meteorites may have killed animals. On May 1, 1860, for instance, a thundering bolide exploded over New Concord, Ohio. The blast rained down at least 30 meteorite fragments [with a total weight of 200 kg (500 pounds)] across an area some 3 × 10 miles; one of these pieces reportedly killed a calf. And on June 28, 1911, a detonating bolide released about 40 stones that fell in the Nakhla region of Alexandria, Egypt, one of which apparently killed a dog; interestingly, not only did the Nakhla stones hail from Mars, but researchers found in them evidence supporting the theory that life once existed on that planet.

The first well-documented strike on a human occurred on November 30, 1954, in Sylacauga, Alabama — in a home across the street from the Comet Drive-in Theater! That day Ann E. Hodges (1923–1972) was taking a midday nap on her couch when a 1.4-kg (3-lb), grapefruit-sized meteorite crashed through the roof of her house, bounced off a console radio, then smashed into her hip. Fortunately, the only injury she sustained was a swollen and painful bruise.[2]

Meteorites have also had other curious effects on humans. On September 15, 2007, a 3-m-wide (10-foot-wide) stony meteorite crashed near the village of Carancas, some 800 miles (1,300 km) south of Lima, Peru. Because of the high location of the site, this was the very first time in known history that the rock was not completely slowed down from its cosmic speed by our atmosphere. The object impacted the Earth at an estimated speed of 24,150 km (15,000 miles) per hour, leaving a crater some 15 meters (50 feet) wide and 6 meters (20 ft) deep. The impact ejected soil nearly the length of four football fields away. When locals went to investigate the impact site, 600 of them succumbed to mysterious "headaches, vomiting and nausea." Actually, the maladies

[2] Hodges became an instant celebrity as the only person on Earth to have been hit by a stone from outer space. According to John C. Hall of the University of West Alabama, the emotional impact of her brush with celebrity, along with other factors, contributed to her and her husband separating 10 years later. "They both agreed," Hall writes in the *Encyclopedia of Alabama*, "that the emotional impact and disruption caused by the meteorite were contributing factors and said they wished it had never happened."

weren't mysterious at all. Researchers now believe that the meteorite most likely vaporized arsenic-containing water that was near the surface of the impact site, and onlookers and investigators breathed in the noxious gas.

The art of meteor observing

Catching sight of a slow-moving fireball through your binoculars is a task that requires you either be looking at the sky with binoculars in hand when the fireball appears, or that the fireball burns slowly enough for you to grab your binoculars (if they're nearby). Otherwise, average meteors usually come and go too quickly for you to catch.

On most nights, meteors are infrequent and random. We have no way to know beforehand when and where to point our binoculars for an encounter. Certainly, if you spend enough time sweeping the heavens with your binoculars you'll eventually catch one of these random,

or *sporadic*, meteors streaking through your field of view. Sometimes the apparitions can be quite interesting or curious. I have caught twin meteors, diffuse meteors, and tumbling meteors. Just expect the unexpected.

It's also true that you can increase your chances of catching a sporadic meteor in your binoculars by observing the sky between midnight and dawn. Around midnight, your observing site begins rotating toward daybreak, in the direction that the Earth is facing as it orbits the Sun − the direction in which Earth is plowing headlong into interplanetary debris (see the top diagram on page 124). Just as bugs more frequently *splatter* onto a car's front windshield than onto its rear as the car races forward, so too meteors can appear more frequently between midnight and dawn than from dusk to midnight, when we're looking out Earth's "rear window."

Meteor showers

But the best way to observe meteors are during annual *meteor showers*. Unlike with sporadic-meteor activity, meteor showers are predictable, occurring on or around the same dates each year. Meteor showers occur whenever the Earth passes through a *meteoroid stream* − a trail of solid particles released from a periodic comet or an asteroid − that intersects Earth's orbit. During the passage, we see a burst of meteor activity radiating from a single point on the star sphere (the *radiant*).

The radiant, however, is a visual illusion. Meteors strike the Earth in parallel paths, but perspective causes us to see them fanning out radially from a single point in the sky. It's the same illusion that causes parallel railroad tracks to appear to converge on a single point on the distant horizon, or the parallel rays from the Sun appear as sunbeams radiating from the Sun in the sky (see the bottom illustrations on page 124).

The duration of a shower depends largely on the width of the stream. Young meteoroid streams are narrow enough for Earth to voyage through them in just a few days; older streams are much wider, so it may take Earth several weeks to clear the debris field. And while some showers may persist for weeks, their peak activity generally lasts less than a day. Some showers, though, are noted for having secondary and tertiary peaks that can spread out maximum activity for a few days.

Most major showers generate meteors at a rate of one meteor every few minutes. But periodic elevated peaks (100+/hour) are common. On rare occasions, shower meteors may fall so fast and furious (several per second) that the night sky will look as if it's weeping starlight. Patient observers who raise their binoculars to the sky during any shower maximum . . . and wait . . . will maximize their chances of seeing meteors come *crashing* into view.

But before you start your adventure, you should be fully prepared for your watch and know how best to record what you see. Here are some simple basics. You can expand on these and be as imaginative as you want. You may also want to join a meteor observing group so that you can share your reports.

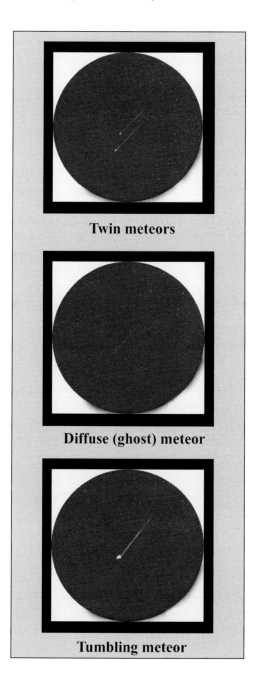

Twin meteors

Diffuse (ghost) meteor

Tumbling meteor

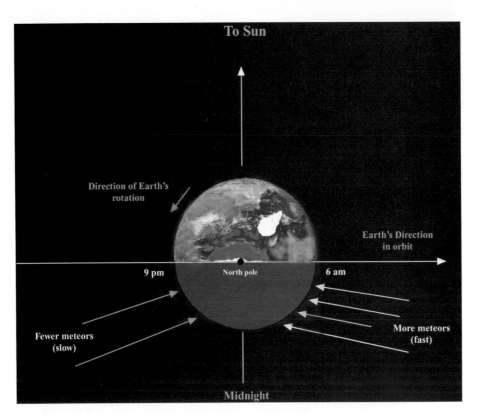

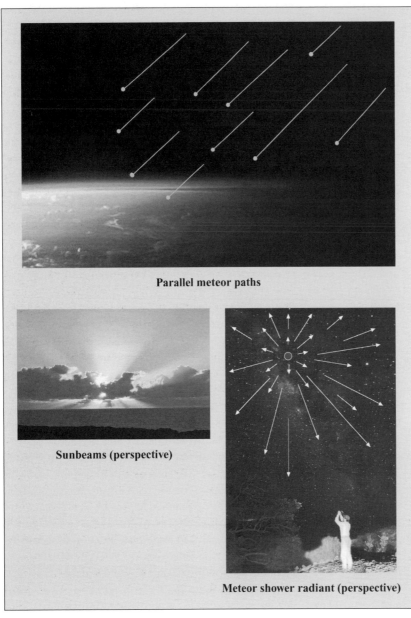

Parallel meteor paths

Sunbeams (perspective)

Meteor shower radiant (perspective)

Shower name	Period of activity	Dates of max. activity (UT)	Radiant at max. (R. A.)	Radiant at max. (Dec.)	Velocity (km/sec)	ZHR	r	Comments
				Some trustworthy meteor showers				
Quadrantids	Jan. 1–5	Jan. 3–4	15.3^h	+53°	41	60–130+	2.1	Fast; aqua streaks
April Lyrids	Apr. 16–25	Apr. 22–23	18.2^h	+33°	47	12–100+	2.9	Fast; white; trains
η Aquarids	Apr. 18–May 28	May 7–8	22.6^h	−0°	66	28–85	2.4	V. fast; long paths
Scorpiids								
Alpha	Apr. 21–May 22	May 16	16.7^h	−23°	35	3–5	2.5	Fireballs; daytime
Omega	May 23–June 15	June 2–3	16.3^h	−18.5°	23	3–5	3.0	Fireballs; daytime
S. δ Aquarids	July 12–Aug. 19	July 28–29	22.8^h	−16°	40.5	18–20	3.2	Nice for binoculars
Perseids	July 17–Aug. 24	Aug. 12–13	03.1^h	+58°	60	23–100+	2.6	V. swift; rich
Draconids	Oct. 6–10	Oct. 9–10	17.5^h	+56°	20.5	~20–500+	2.6	V. slow; per. storms
Orionids	Oct. 2–Nov. 7	Oct. 21	6.3^h	+16°	66	25–35	2.5	V. fast; long trains
Leonids	Nov. 14–21	Nov. 17	10.2^h	+22°	71	10–100+	2.9	V. fast; v. long trains
Geminids	Dec. 7–17	Dec. 13–14	7.6^h	+32°	35	18–120	2.6	Fast; yellow fireballs

Plan ahead

Some trustworthy meteor showers as listed on page 125. (Detailed descriptions of these events begin on page 129.) The calendar is based largely on data from the International Astronomical Union's Meteor Data Center (dates of maximum activity, radiant position, and meteor velocities) and the International Meteor Organization's Working List of Visual Meteor Showers (period of activity, and r values). The zenithal hourly rate (ZHR) values reflect data from both organizations. The comments at the end are based largely on my own observations. The dates of maximum activity are given in Universal Time (UT).

The first column gives the *shower's name*. Meteor showers are usually named after the constellation in which their radiant appears to lie – a convention the great Italian observer Giovanni Schiaparelli introduced in the 1860s. The Perseids, for example, have a radiant in Perseus the Hero, and the Orionids in Orion the Hunter. Some showers may have multiple radiants or radiants that occur in the same constellation but at different times of the year. In these cases, the shower may be named for the constellation and the month in which the shower occurs or the brightest star near the radiant.

The April Lyrid meteor shower, for instance, is a major shower with a radiant in Lyra the Harp. Its activity peaks in April, as opposed to the minor shower, the Eta Lyrid meteor shower (not listed) with a maximum in May. Sometimes a shower has multiple radiants in the same constellation. The Delta (δ) Aquarids, for example, have a double radiant with north and south components; the Southern Delta Aquarid radiant produces the more prominent meteors. The Quadrantids are named for the now defunct constellation Quadrans Muralis, the Mural Quadrant. But there are other exceptions to Schiaparelli's rule. The Draconids, for instance, have a radiant in Draco the Dragon, but are also popularly known as the Giacobinids – in honor of their parent comet, Giacobini–Zinner.

The second and third columns show the *period of activity* (which reflects the time it takes for Earth to pass through the meteoroid stream) and the *dates of maximum activity*

(when Earth is expected to pass through the densest part of the meteoroid stream, giving us the best show). Neither of these are absolutes, so be flexible. Each year, the International Meteor Organization (IMO) publishes predictions for shower activity, based on recent research, and is an excellent resource (http://www.imo.net/calendar).

The fourth column gives the coordinates [*right ascension* (R. A.) and *declination* (Dec.)] for the *radiant at the time of maximum* activity. These data were used to plot the radiants shown in the photo illustrations that accompany the detailed shower descriptions. You can also use the data to plot them yourself.

The fifth column, *velocity*, is the atmospheric or apparent meteoric velocity, given in kilometers per second. The velocities in this list range from about 20.5 km/sec (slow) to 71 km/sec (exceptionally fast). This information is especially helpful to the binocular user, because the slower the meteor, the better your chances of nabbing one in your binoculars.

The ZHR is a calculated maximum number of meteors a skilled observer would expect to see in one hour under a perfectly dark and clear sky with the shower radiant overhead. It's extremely important to accept that a shower's period of activity, dates of maximum activity, and ZHR are not set in stone, and you may find little or great variation in them. Zenithal hourly rates are especially susceptible to high swings in values. Not everyone is under an ideal sky (naked-eye-limiting magnitude 6.5), and radiants hardly ever get to the zenith for most observers. The lower the radiant is in the sky, the fewer meteors you will see. As a general guideline, the IMO estimates that if a shower has an impressive ZHR of 100, but the radiant at your site is only 40° above the horizon, an observer should expect a ZHR of 64; if the radiant is only 10° above the horizon, the ZHR drops to 17. Of course, interference from a bright Moon or artificial lights will diminish the number of faint meteors you will see. To calculate the expected shower rate at a given date and location, go to http://leonid.arc.nasa.gov/estimator.html.

But, as you will see in the descriptions of these showers to follow, surges in ZHR are not uncommon. So expect the unexpected! The ZHRs will also vary greatly depending on the exact time of predicted maximum; for instance, if a shower puts on its strongest display between the hours of 3 and 4 Universal Time, and darkness does not fall on your location until several hours later, your ZHR may be much lower than that for observers fortunate enough to be in the right place at the right time.

The parameter r is the population index, a value you will need to know if you want to calculate the ZHR at your location at the time you were observing (see page 128). Loosely, the value reflects the brightness distribution of meteors in the shower. An r value between 2.0 and 2.5, for instance, means that the meteors will be brighter than average (so you can expect a greater than average showing of meteors brighter than 2nd or 3rd magnitude). An r value greater than 3.0 means that the shower produces many faint meteors. The average r in my selected list is 2.7.

Finally, the *comments* are a quick, at-a-glance summary of what to expect based largely on my own observations of the showers over the years. Again, all the major showers are described in greater detail beginning on page 129.

Plan your observation carefully. Under ideal circumstances, you'll get a chance to observe a meteor shower when its radiant is highest in the sky around the time of predicted maximum activity – especially if that occurs in the early morning hours before dawn – with no Moon in the sky. Most of the time, however, you'll just have to play with whatever cards Lady Chance gives to you.

But you do have some control. To maximize your chances of seeing the widest distribution of shower members (from bright to faint), and therefore increasing the number of meteors you'll see, you'll want to be in a location that is as far away from city lights as possible, under the clearest and most transparent skies as possible, and be looking during a time when the Moon is not hideously bright.

Your site should also offer you a full view of the sky with little or no horizon obstructions, such as tall trees. And while you certainly can't remove the Moon from the sky if it happens to be full around the time of a meteor shower's activity, don't despair. As you will discover in my detailed shower descriptions, some events produce bright fireballs that can be seen even during full moonlight and still be spectacular.

Gearing up

To maximize your pleasure, and your chances of seeing meteors, you'll want to be out under the stars for at least an hour. And since you'll be looking up at the sky while largely motionless, you'll want to be comfortable. Remember, the two words skywatchers perceive as both a blessing and a curse are "clear and cold." So dress warmly enough for the season, keeping in mind that even summer nights can bring a chill.

Lay on a blanket (which you can also use as a wraparound) or a sleeping bag, or bring a folding chair or recliner and drape a blanket over you. If you're lying down, prop your head up with a pillow to alleviate neck strain, especially for those times when you're using binoculars. Bring a cooler full of food, snacks, and fluids; warm drinks, like hot chocolate from a thermos, sure helps on those cold nights. It's your watch, so bring along whatever you need to be in complete comfort.

As for observing equipment, you'll want a watch, a flashlight with a red filter,[3] binoculars, star wheel, pens or pencils, a pocket pencil sharpener, and a note pad.

Some people prefer to record their observations with a tape recorder; the choice is yours. My suggestions below for recording meteor data are for those using notepads, which you should bring as a backup, regardless, because mechanical devices can fail and frustrate, while a pencil and paper is virtually flawless. A small collapsable table is also very handy, but optional. Plan to arrive at the observing site at least a half hour before you start your watch; you'll need time to set up and to determine the site's limiting magnitude. The darker the sky, the fainter you'll see. If you can see to magnitude 6.5 or fainter with your unaided eyes, you're in an ideal location. Under a decent suburban site, a naked-eye observer can see to magnitude 5.8 at best. A site that has a naked-eye limit of 5.3 magnitude is generally about as good as a dark sky under full moonlight.

You'll also want to get a feel for how bright stars appear in your binoculars. This will help you to estimate the brightness of binocular meteors. Knowing how faint you can see while looking casually through your binoculars will help you place a lower limit on the brightness of the

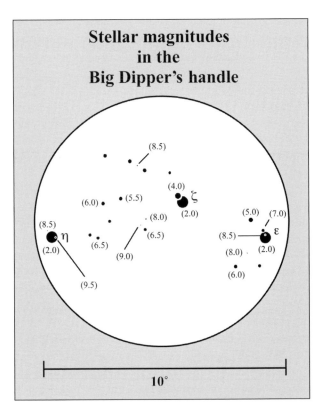

Stellar magnitudes in the Big Dipper's handle

10°

---

[3] Your eyes are insensitive to red light; white light bleaches out the night-sensitive rod cells in your eyes, diminishing your chances of seeing faint meteors.

meteors you're recording. Under a very dark sky a casual glance through 10 × 50 binoculars will show stars around magnitude 8.5, meaning most meteors you'll notice will be around 8th magnitude or brighter. Use the star chart below of stars in and around the Big Dipper's handle to find your naked-eye and binocular limits.

Once you're set up, select a region of sky to concentrate your attention with your unaided eyes. The radiant is not the best place. Most shower meteors appear some distance from that spot; it's just that you can visually trace them back to that point of origin.

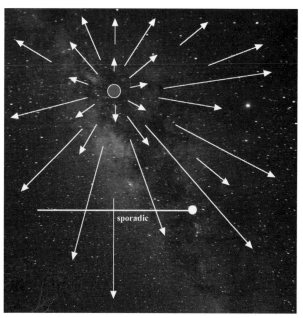

But that doesn't mean you *have* to avoid the radiant, which is the center of an area of sky slightly smaller than your fist held at arm's length. It's just that the meteors seen here are more difficult to detect, because they're commonly short streaks of light. But one benefit to looking at the radiant is catching a *point meteor* – a point of light that rapidly flares and dims before your eyes; believe me, it's an eerie sensation seeing a cosmic rock zeroing in on you . . . especially if it's bright!

The meteors that appear farther away from the radiant have longer paths and are more likely to attract attention. It's recommended to focus your attention about 20° to 40° away from the radiant, where, statistically, most meteors are seen. It's also beneficial to look about half way up the sky, because you'll be able to cover a greater area of sky than if you were looking directly overhead. But the sky is yours. So you decide. Wherever you choose to look, you will need to identify the radiant's position so that you can differentiate shower meteors from sporadic meteors. All shower meteors can be traced back to the radiant; sporadic meteors go against that visual grain.

What to record

Everyone has their own method of meteor recording. If you join a meteor organization you'll have to follow their recommendations and rules.

The easiest method is what the IMO calls the "counting method," in which the observer simply uses a dig-

ital recorder or notepad to keep track of the number of meteors detected during the observing period, noting the magnitude and color of each meteor, as well as if it was a shower meteor or not. The other method is to plot each meteor on a star chart. Plotting is a fine method, but only if the shower is "slow," because plotting requires time away from monitoring the sky to draw each meteor's path. The more time you spend looking at a star chart, the less time you have to see meteors, so your counts may be unreliable.

Plotting is extremely helpful when several meteor showers are occurring around the same time (say one major, and two minor shows). In this case, the IMO recommends combining the counting and plotting methods:

> All meteors which could belong to one of the minor showers are plotted while obvious Orionids and Sporadics are "counted" according to the guidelines for major-shower observations, i.e. you record the latter meteor data onto tape (or write the details as notes) without taking your eyes off the sky, while plotting the former meteors. In this way you reduce the amount of dead time but still enable an accurate shower association to be obtained for the minor showers.

> *As soon as you see more than 20 meteors per hour you should plot only meteors which could belong to a minor shower; other meteors are "counted" only.*

It's important to consider when a shower peaks, and when during the shower's predicted activity cycle you are observing. Major showers can be considered minor showers if you are observing before or after their predicted period of peak activity, because the rates are so low. Choose your recording method appropriately. The IMO recommends you plot all meteors seen if you're seeing less than 20 meteors per hour. If there's more than one shower occurring, and one of them is quite active, producing 50 meteors or more per hour, focus most of your attention on that active shower.

If you want to remain an independent observer, and if the meteor shower is strong (with meteors occurring rapidly), you might consider adopting my own personal method, which has served me well for several major showers.

Once you're ready to observe, start by recording the meteor shower, the date, your location, your magnitude limits, and any other pertinent information, such as cloud cover. Then begin your watch by recording the starting time. As you will most likely be seeing meteors at a rate of one every few minutes, with occasional spurts of a few meteors in a matter of seconds, you'll have time to take decent notes after each fall. During dramatic meteor showers I do not record the time of every meteor, unless it's a fabulous fireball or bolide. Rather, I prefer to record the time every 10 to 15 minutes or so (depending on the activity) until an hour is up. Then I start again. That way I don't have to keep looking at my watch instead of the sky.

In between the time records, I record meteor falls using my own method of notation. Here is a hypothetical sample:

In this example, the number is the estimated magnitude of the meteor. The line (—) following it represents the perceived path (short, medium or long). The line with the circle at the end (——ø) shows that the meteor had a bulbous head. The lower case letter after the line records the color (g = green; b = blue; y = yellow); I don't record the color if the meteor is white. The × represents an exploding meteor (in this case a magnitude −5 fireball). The wavy line (∼∼∼∼∼) represents that it left behind a train. And the capital S stands for a sporadic meteor. Periodically, during lulls in activity, I make notes about the event.

As you can imagine, with practice, it doesn't take long to jot down, say, a 4 and a short line and a few other letters or symbols. I've gotten to the point that I can make these records sometimes without taking my eye from the sky. For instance, recording that −5 fireball would have taken me less than 5 seconds. Note that in the second hour I switched to binoculars and just recorded the magnitudes and colors. The P stands for point meteor.

If the shower is very intense, counts are most important, so I just record the start time, then use my pen or pencil to notch the notepad one notch for every meteor. If there's a lull in activity, I record the time, then make quick notes.

Quadrantid meteor shower
January 4, 2010
(Kilauea Volcano summit), altitude =
4,200 feet
Naked-eye limiting magnitude = 6.5
Casual binocular limit = 9.5
Skies clear and transparent / No clouds

13:00 UT
(Naked-eye)
4 -
3 ——ø g
2 ———ø b
4 ——
4 -
- 5 ————ø y × ∼∼∼∼∼ (13:20 UT) 0 ——ø
S
4 -
2 ——ø
S
(Many of the meteors have large bulbous heads with thin "rat tails")
14:00 UT
(10 × 50 Binoculars)
Looking at radiant
4
7
P
5 y
15:00 UT
END

Draconid meteor shower
October 10, 2010
(Kilauea Volcano summit), altitude =
4,200 feet
Naked-eye limiting magnitude = 6.5
Casual binocular limit = 9.5
Skies clear and transparent / no clouds

16:00 UT
/// /// // / // / /// // /// // /// /// ///
16:20
// // /// / // / /// /// // /// // / / ///
(Most meteors 3rd magnitude and fainter, an occasional fireball, many come in waves of two or three)
16: 45
// / / / / / // / / / /
(Meteor activity appears to be slowing down)
/// / / / / / // / /
(Activity picked up again briefly before falling off rapidly)
17:00
END

Calculating your own ZHR

Let's say you observed the Leonid meteor shower for one hour on the morning of November 17. The sky was perfectly clear (no clouds), the limiting naked-eye magnitude was 5.0, the radiant had an average elevation of 50°, and you counted 27 meteors in that hour period. To find your ZHR, you'd perform the following calculations:

(1) Calculate $r^{6.5-LM}$, where r is the population index of the Leonids given in the table on page 125 (2.9), and LM is the limiting naked-eye magnitude during your observation (5.0). This correction factor will adjust your tally by considering your sky conditions.

$$2.9^{1.5} = 4.9.$$

(2) Calculate $1/\sin E$, where E is the mean altitude of the radiant during your observation (50°). This value will correct for any difference in altitude from the perfect radiant position at the zenith.

$$1/0.77 = 1.3.$$

(3) Calculate $ZHR = r^{6.5-LM} \times 1/\sin E \times T_m$, where T_m is your meteor tally for the hour (27).

$$ZHR = 4.9 \times 1.3 \times 27$$
$$ZHR = 172.$$

Had you been under perfect conditions, the ZHR for the Leonid shower you witnessed would have been an exceptional sight! You perform these calculations for each hour you observe. For instance, if you continued to observe for another hour and tallied 10 meteors in that hour, it's clear that though the radiant would have risen higher (to ~65°) your tally was lower than in the first hour, reflecting diminishing activity:

$$ZHR = 4.9 \times 1/\sin 65 \times 10$$
$$= 4.9 \times 1.1 \times 10 = 54.$$

Indeed, during the second hour of meteor watching, the ZHR value from your location dropped to 54 from 172 – a decrease in meteor production by nearly 70 percent! One could theorize, then, that in this hypothetical situation you had caught the tail end of a Leonid maximum.

By the way, if clouds interfere with part of your watch, you can correct for that variable by using the equation $1/(1 - x)$, where x is the cloud cover expressed as a decimal. You then multiply that number by your final ZHR. Let's say, for example that the sky was 15 percent cloud covered on average during your second hour of observing. Your revised ZHR, then, would be

$$ZHR = 1/(1 - 0.15) \times 54 = 1.2 \times 54 = 64.8.$$

The error in the ZHR value is $ZHR/\sqrt{N}$, with N the number of observed meteors.

Some trustworthy meteor showers

The following selected meteor showers are those that I feel put on the best and most dependable shows every year. I tell you a bit about each shower's history and give examples of them at their best, sometimes from personal experience. But remember, some showers may be excellent over certain latitudes and poor over others. Some showers are better in certain years than in others. The anticipation should help to excite you. But if the shower lets you down, well, try again the next year. The solar longitude (L_s) given after the name of each shower is a precise measure of the Earth's position on its orbit, which is not dependent on the vagaries of the calendar. All are given for the equinox 2000.0.

Quadrantids (Bootids) ($L_S = 283.3°$)

One great way to ring in the new year is to look for celestial fireworks from the Quadrantid meteor shower. Active from January 1–5, it's been one of the sky's "old faithfuls" for more than 175 years. I rank its peak activity as being tied with that from the Geminids as the year's best; both showers have been consistently rich and dependable. For many viewers living at mid-northern and higher latitudes, the Quadrantid radiant is circumpolar (meaning it never sets below the horizon), though visual observations are usually most plentiful during the wee hours of the morning, when the radiant is highest in a dark sky.

The Quadrantids have a fascinating history. Its name is derived from the now defunct constellation: Quadrans Muralis, the Mural Quadrant – an inconspicuous aggregation of dim stars that filled a void between Draco the Dragon, Hercules the Strongman, and Bootes the Herdsman. French astronomer Joseph Jerome LaLande (1732–1807) created the constellation in 1795 to honor the instrument that he and his nephew used to observe the stars, and it can be found on certain old star atlases,

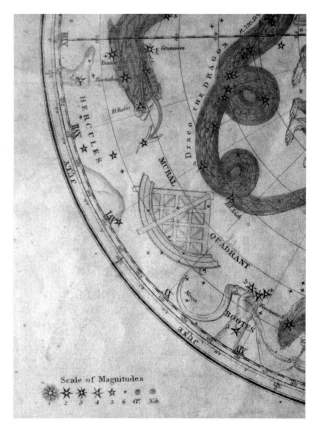

such as Burrit's of 1835. The Quadrantid radiant was located in this region of sky when Louis Francois Wartmann of Switzerland discovered it in 1835.[4] Still, the shower was not recognized as an annual event until 1839, when Adolphe Quételet (Brussels Observatory, Belgium) claimed first publication. Then, in 1922, the International Astronomical Union removed Quadrans Muralis (and several other constellations) from its list of 88 officially recognized modern constellations (see Appendix A in my companion book, *Observing the Night Sky with Binoculars*.) With the constellation boundaries redefined, the radiant now fell within the confines of Bootes, so the shower is sometimes referred to as the Bootids.

Today, peak activity usually occurs around January 3–4 UT, at solar longitude 283.3°. For this shower, it's important to know the predicted time of maximum, because its duration is brief, lasting only a few hours or so. But the visual display during this time can be intense, with zenithal hourly rates commonly exceeding 100. Outside this narrow window, the meteors can be sparse.

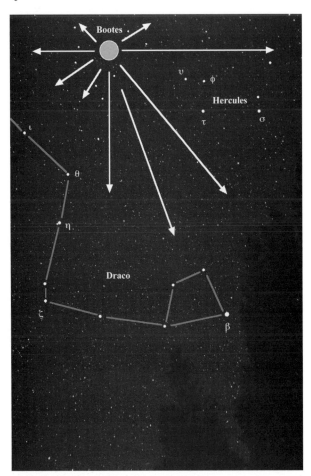

Then again, in a 1988 article in the *Journal of the International Meteor Organization* (volume 16(2), pages 58–61), Bernhard Koch reports some interesting binocular results

[in 1987] in southern France by the youth group of the Astronomische Arbeitsgruppe Ulm. In addition to naked-eye counts, the expedition members made observations with different binoculars at different distances from the radiant and found that the binocular maximum occurred 24 hours before the naked-eye one. Peter Jenniskens, a meteor astronomer with NASA Ames and the SETI Institute, adds that underlaying the main peak is a structure rich in faint meteors that peaks a day earlier but is much wider.

Quadrantid meteors are swift. Fireballs (−5 magnitude or brighter) are rare, with the vast majority of meteors ranging in brightness between 0 and 4. My observations show that the brighter ones often appear lemon in color; they also have large bulbous heads with thin "rat tails." The fainter meteors commonly come in two varieties: fuzzy, aqua teardrops; and squat reddish teardrops when seen at lower elevations (usually when the radiant is just rising and the meteor is low and paralleling the horizon). Because of their speed and lack of brilliant members, it's hard to catch one leaving a persistent train.

But I have found it common to see, in an hour, some Quadrantids split during flight or appear in faint fuzzy pairs. If you're lucky enough to raise your binoculars to the right position in the sky when this happens, you may be in for a visual treat. The photo illustration below shows the great variety in appearance of the Quandrantids, seen with the radiant low to the horizon.

---

[4] Gary Kronk (http://meteorshowersonline.com) notes that Antonio Brucalassi may have been the first to document seeing a meteor shower on this date; on the morning of January 2, 1825, Brucalassi saw the "atmosphere... traversed by a multitude of the luminous bodies known by the name of falling stars."

For many years, the Quadrantids was almost unique among the major meteor showers, because its parent body had not been clearly identified. That changed in 2003 when Jenniskens identified the parent body as an asteroid-looking object called 2003 EH1. The asteroid had been discovered months earlier by Brian Skiff, who used the Lowell Observatory Near-Earth Object Survey (LONEOS) telescope. Jenniskens found that when 2003 EH1 is in the same steeply inclined orbit and when it passes near Earth's orbit, it's located precisely at the peak of the Quadrantid stream. Furthermore, it appears that asteroid 2003 EH1 may be a fragment of Comet C/1490 Y1's nucleus. Indeed, Japanese astronomer Ishiro Hasegawa calculated a parabolic orbit for Comet C/1490 Y1 and pointed out the similarity with the orbit of the Quadrantids. The Quadrantids, then, is a relatively young meteor stream being on the age of 500 years, which helps to explain its brief but intense maximum.

Jenniskens predicts that this shower will remain strong over the next 100 years, but it will also vary considerably in peak intensity from year to year due to Jupiter's perturbations at aphelion.

April Lyrids ($L_S = 32.4°$)

With the cold of winter on the wane, the April Lyrids are a refreshing start to springtime meteor observing. The shower is an exciting event, in that one never knows what to expect. The Lyrids are, in fact, one of the more mysterious major meteor showers and well worth watching. While you can expect to see at least about 15 to 20 meteors per hour at maximum (especially when the radiant is high above the horizon before dawn), the shower occasionally exhibits intense bursts of activity.

According to Biot's *Chinese Catalogue*, the April Lyrids can be traced back to 687 BC, when "in the middle of the night, stars fell like rain." The phenomenon repeated itself in 15 BC, when "after the middle of the night, stars fell like a rain; they were 10° to 20° long; this phenomenon was repeated continually. Before arriving at the earth they were extinguished."

As witnessed from the eastern seaboard of the United States, the Lyrids made modern history in 1803, when the shower was nothing short of spectacular. In his 1925 book, *Meteors* (Williams & Wilkins Company; Baltimore), Charles Olivier includes an April 23, 1803 newspaper account appearing under the heading "Shooting stars" in the *Virginia Gazette* of Richmond.

This electrical phenomenon was observed on Wednesday morning last at Richmond and its vicinity, in a manner that alarmed many, and astonished every person who beheld it. From one until three in the morning, those starry meteors seemed to fall from every point in the heavens, in such numbers as to resemble a shower of sky rockets. Several of these shooting meteors were accompanied with a train of fire, that illuminated the sky for a considerable distance. One, in particular, appeared to fall from the zenith, of the apparent size of a ball of 18 inches [diameter], that lighted for several seconds the whole hemisphere. During the continuance of this remarkable phenomenon, a hissing noise was plainly heard, and several reports resembling the discharge of a pistol.

The estimated ZHR for the 1803 shower is 860. The Lyrids had a similar outburst in 1922, a weaker one in 1945 (ZHR = 47), and a moderately strong showing (ZHR = 250) in 1982. While the observed outbursts occur at infrequent intervals (some outbursts may have been missed owing to poor weather), the intervals between the outbursts appears to be in multiples of 12 years.

In most cases, such meteor-shower outbursts occur at time intervals that closely coincide with the period of the parent comet. For instance, as we will see, November's Leonid meteor shower tends to have extraordinary displays roughly every 33.5 years, which coincides with the period of its parent comet: 55P/Tempel–Tuttle. But the parent comet of the April Lyrids is C/1861 G1 (Thatcher), which has an orbital period of 415 years – far in excess of 12 years. What then is causing the outbursts? In a 1995 article in *Monthly Notices of the Royal Astronomical Society* (volume 277, pages 1087–1096), Terence R. Arter and Iwan P. Williams (University of London), investigated

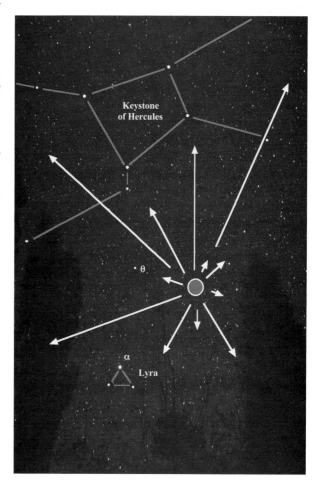

various hypotheses for the origin of the outbursts of activity (including a breakup of the comet's nucleus into a multitude of discrete fragments sometime in the past) and found no convincing explanation for the observed periodicity.

That same year, was Peter Jenniskens able to prove that this periodicity is on account of a narrow dust trail moving in and out of Earth's path, following the wagging Sun's motion in response to Jupiter's 12-year orbit.

In a 2001 article in the *Journal of the International Meteor Organization* (volume 29, pages 119–133), Audrius Dubietis and Rainer Arlt made the most detailed examination of the shower in modern times, finding that the time of maximum varies from year to year as does the peak's length; for instance, it varied from 15 hours in 1993 to 62 hours in 2000, with a mean value of 32 hours. Still, the most intense period of activity during that peak interval may only be a few hours. Even the annual shower appears to move in and out of Earth's path.

Still, it pays to watch this shower on the days before and after the predicted maximum on April 22–23 UT. The radiant is about 7° (or one binocular field) southwest of brilliant Alpha (α) Lyrae (Vega). Even on an average year, the shower can produce some nice bright fireballs that leave persistent trains. This makes Lyrid observing great for binocular users. The long flight paths of the meteors may also give you enough time to train your binoculars on the object as it flies across the heavens in blazing wonder. To make the most of the shower, though, especially on Moonless nights around the times of maximum, you may want to be situated under dark skies; Dubietis and Arlt also found that the Lyrid peak is occasionally accompanied by a short-lived increase of fainter meteors. Of course, the fireballs can be enjoyed from suburban or city locations.

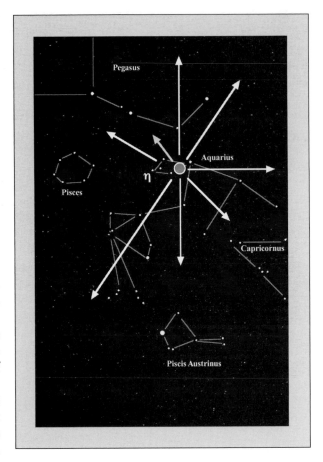

Eta (η) Aquarids ($L_S = 46.9°$)

Many youngsters are introduced to meteor showers during the August Perseids; school is out, the nights are relatively warm, and the shower's radiant is up as darkness begins. But my first experience was in early May with the Eta (η) Aquarids. In the mid 1960s, I had learned that these meteors were phantom fragments of Comet 1P/Halley, the most popular comet in history. From my childhood home in Cambridge, Massachusetts, the shower's radiant (near the water jar of Aquarius) is highest just before dawn; even then it is relatively low in the southeast. Still I planned to see them. So a friend and I created a blanket tent on our joint back porch and waited until the post-midnight hours in our sleeping bags for the radiant to rise.

At first the meteors appeared as little whispers of light. But they soon became quite swift and bright, traveling across the sky in long and slender paths. Throughout the night, the activity repeated in waves – with one or more meteors appearing in a minute, followed by several minutes of quiet before the action picked up again. As each fleeting spirit materialized, I could envision the Earth racing through space and smashing head-on with these dust particles ejected from the great comet's sweaty head, only to see them burn up and vanish from view high in Earth's atmosphere – though some left behind thin glowing trains, which we could see gently twisting over time through our binoculars. Seeing the heavens weep starlight was one of the many significant moments in my youth that helped to keep the flame of astronomy burning in my heart.

The Eta Aquarid meteors are indeed very swift, penetrating the atmosphere at speeds of 66 km per second (41 miles per second). That's because Comet Halley (and thus the Eta Aquarid meteors produced by it) move in retrograde orbits, meaning they travel opposite to the way Earth rounds the Sun. As a result, each May, the Eta Aquarid meteors collide nearly head-on with the Earth. These high-energy impacts are responsible not only for the great speed in which they enter the atmosphere, but also for the high percentage ($\sim$25%) of long-duration trains left behind by these "shooting stars."

Actually, Earth intersects Comet Halley's orbit twice each year: once near the comet's descending node in April/May (producing the Eta Aquarid meteor shower, which peaks around May 7–8 UT), then again near the comet's ascending node in October/November (producing the Orionid meter shower [see page 138], which peaks around October 21).

Eta Aquarid displays have been noted as far back as 74 BC. The shower consists of two components, the first

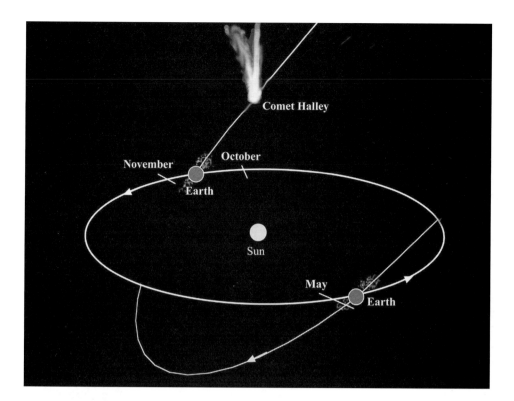

and strongest of which is variable from year to year. The meteor stream is relatively broad, so decent activity can be seen for many days. The stream also has a variable number of sub maxima. Generally, the ZHRs hover around 30 between about May 3–10. Fresh data from the IMO (collected between 1984 and 2001) shows that the peak rates appear to be variable on a roughly 12-year time scale.

Peter Jenniskens has shown that the stream was created ever since the moment that Jupiter captured Comet Halley in a shorter orbit about 18,000 years ago. The comet orbit has since evolved from passing near Jupiter to passing just inside Earth's orbit.

Scorpiids

Alpha Scorpiids ($L_s = 55.9°$)
Omega Scorpiid Complex ($L_s = 71.9°$)

One of my favorites, the Scorpiid meteor shower is not a rich one. But what it lacks in numbers, it sure makes up for in punch. The shower produces only about three to five meteors per hour at their zenith, but it can certainly be an experience to see one. Indeed, the Scorpiids are best known for their eye-catching fireballs that exceed the brilliance of Venus. These objects are bright enough to be seen in twilight or daylight, especially when the Sun is near setting.

The Scorpiids are active from late April to mid June. But the meteors come from three different radiants: The Alpha (α) Scorpiids (which peak around May 16 UT), and the Omega (ω) Scorpiid Complex, which has two radiants (north and south) most active near June 2 UT. One of

my first recorded fireballs was on May 14, 1970, and the meteor flew right out of the Alpha Scorpiid radiant. It was bright enough, and remarkable enough, for my 13-year self to chart its position on a special hand-drafted star chart, which I had created as a way to document transitory celestial events such as meteors and artificial satellites.

The Alpha Scorpiid radiant is about 4.5° northeast of 1st-magntiude Antares, the blood-red heart of Scorpius the Scorpion. For observers at mid-northern latitudes, this region of sky just peeks above the eastern horizon around the time of sunset in mid May – the perfect setting for Earth-skimming fireballs, which overtake our planet from behind. Because of the glancing angle, it's common to see brilliant fireballs spitting debris as they flame across the twilight (or daylight) sky. Peter Jenniskens, a meteor expert with the Carl Sagan Center at the SETI Institute, says that the Alpha Scorpiids move in very elongated orbits, with a perihelion only around 0.2 astronomical units and an aphelion near Jupiter's orbit. Consequently, the meteors are relatively fast, entering Earth's atmosphere at speeds of 34 km (21 miles) per second, making them about as fast as a Geminid (see page 142).

The Omega Scorpiid complex is another fireball shower that immediately follows the Alpha Scorpiids, from May 23 to June 15, with a peak occurring near June 2–3 UT. The mean center of activity lies about 8° northwest of Antares and just west of Chi (χ) Ophiuchi. Unlike the brisk Alpha Scorpiids, however, the Omega Scorpiids take long and lumbering paths through the atmosphere at speeds of only 12 miles (20 km) per second – twice as slow as an Alpha Scorpiid meteor and almost four times slower than a Leonid.

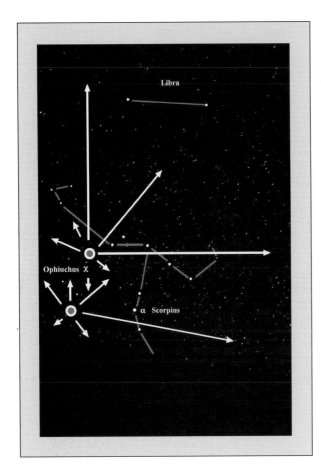

All Scorpiid meteors belong to very broad streams, so you can see activity for weeks before and after the predicted maximum. I call them casual showers, because they require leisurely vigils. If the sky is clear around the time of sunset, I go outside, walk the dog, or just sit down in a chair and keep watching the deepening twilight, at least until the stars come out. If I happen to be out under the stars at night, I'll keep a periodic eye on the naked-eye sky for that rare Scorpiid.

Not many popular sources include the Alpha Scorpiids in their lists of meteor showers, so many of the more exciting reports come from the late nineteenth century, when interest in the shower was at its peak. In an 1884 issue of *The Observatory* (volume 7, pages 136–138), William F. Denning of Bristol, England, (that century's most productive and proficient meteor observer) called the Scorpiid meteor shower an "important" one, noting that the "presence of the Moon is an objection; but neither moonlight nor the strong twilight of this season will materially affect the observation of these brilliant meteors."

Denning goes on to recount several dramatic sightings from the fireball shower. At 8:20 pm on June 7, 1878, for instance, a Mr. G. Holmes (Redland, Bristol) saw a Scorpiid in "broad daylight . . . appearing like a globe of liquid fire, with pale green tinge." A Mr. W. Humpris, who saw the same meteor from Bath, "fancied he heard a crackling sound." That same evening, at about 9:50 pm, a Mr. H. M. Rogers, writing to *The Times*, from Knole Park near Bristol, made this descriptive account:

Last night [June 7] I was walking along a footpath close to Knole Park, my shadow falling on the park palings. Suddenly I saw two shadows, one of which was moving rapidly along the paling, the reverse way to the way I was walking. On turning around I saw the largest and brightest meteor it has ever fallen to my lot to witness. It was passing apparently from S. to N., very nearly parallel to the horizon . . . When I saw it first it was about 5° from the moon . . . It passed slowly along about 3° below the moon, or about 30° above the horizon . . . The light was of a very pale green, as nearly as possible like the light of a glowworm highly intensified. As it passed under the moon its brilliancy caused the moon to look of a muddy yellow colour, like a street lamp in a November fog.

Another correspondent at Clifton commented, "I have seen many meteors, but never anything so large, of such brilliant colour [emerald green], and apparently so close to the earth." Five years later, Denning also recognized the coincidence of three extraordinary fireballs from Scorpio: two in 1908; one on May 19, which "passed over Ireland from Ballyteigne Bay to Co. Mayo . . . along a path of 142 miles," and the other on May 22. The third, on May 19, 1906, was a "large meteor" from the same radiant, seen at Bristol and in Wales. He concluded that meteor observers "will doubtless meet with further interesting instances of these brilliant summer Scorpiids in future years, and will find [they] amply repay attention."

Indeed, in 1908, one June fireball event literally shocked the world. Early on the morning of June 28, an object of extraterrestrial origin rushed through Earth's atmosphere at supersonic speed before it detonated in the air with an energy equivalent of nearly 1,000 Hiroshima-type atomic blasts. The dual shock waves created by those events – one from the object's entry, the other from the blast – slammed into the Earth, laying trees flat across 50 miles of forested swampland in Tunguska, Russia. Seismographs in Washington, D.C., registered the blast, which threw enough dust into the atmosphere to dim the skies over North America for weeks.

In a 1930 letter to *The Observatory* (volume 53, pages 177–178), Denning questioned whether that great fireball was related to the June fireball shower he had earlier detected. Although Denning did not see that event, he did witness its remarkable atmospheric effects: "All the northern region was strongly illuminated – at midnight a game of cricket was played on Durdham Down, Clifton, and various other avocations, only possible in the half-light of an ordinary night, were freely indulged in." The effects continued through July 1, when the general light was less diffused.

"During more than 60 years of night observations," Denning continued, "I cannot recollect seeing the firmament so light as at this period." The great observer went on to note that "ordinary shooting stars were scarce on the various dates mentioned, but on June 28 a fireball was noticed by me at Bristol traversing a long flight from a radiant at [R.A.: 16^h 00^m; Dec. $-18°$ (the Omega Scorpiid Complex)], and on July 1 of the same year a large fireball was abundantly recorded from a nearly identical center . . . Possibly the objects had their origin in the same system as that which supplied the great meteor of 1908 June 30. I have never seen any computation of the real path of the latter object."

Could it be? When you look up at the summer sky and see a slowly burning fireball from Omega Scorpiid Complex, you're seeing a sibling of the Great Siberian Fireball?

Alas, no. Modern analysis shows that the fireball came from the southeast early that morning. Scorpius would have already set in the west. But some astronomers do suspect that the Siberian fireball might have been a small asteroid, roughly 30 m (100 feet) in diameter, which may have been a member of the Beta Taurid meteor stream. This meteor shower is active at this time of year in the daytime; astronomers track the meteors by radio techniques.

To this day, the parent body of the Omega Scorpiid meteors remains elusive, though the Near-Earth Asteroid 2004 BZ74 appears to be the parent of the Alpha Scorpiids. So who knows what excitement the future brings?

Southern Delta (δ) Aquarids ($L_S = 125.6°$)

While you're waiting for the Perseids to culminate around August 12–13, you can keep your eyes peeled for Southern Delta (δ) Aquarid meteors. They're active from July 12 – August 19 and peak around July 28–29. The shower has a respectable ZHR of 20, but if you're an observer living at mid-northern latitudes, you can expect a meteor every 5 to 10 minutes or so around the time of maximum. The meteors are swift but many are relatively dim, making them a good shower for binocular users. Still, enough bright members appear to make the shower worthwhile for the naked-eye observer. International Meteor Organization observations have shown the the time of maximum activity may actually be broader than thought, lasting from July 27–29.

The Southern Delta Aquarid radiant is a few degrees southwest of 4th-magnitude Delta (δ) Aquarii, or a little more than 10° (a fist held at arm's length) north and slightly west of 1st-magnitude Alpha (α) Piscis Austrini (Fomalhaut). Shower activity has been fairly consistent over the years, but some variability has been suspected. The shower's parent comet belongs to the Marsden Group of "sunskirting" comet fragments, which belong to the Machholz complex – objects on currently different orbits but showing the same orbital evolution. They are named for Brian Marsden of the Harvard/Smithsonian Center for Astrophysics who announced in 2004 his discovery of the comet group on Solar and Heliospheric Observatory

(SOHO) images. The Southern Delta Aquariid meteors are in the same orbit as these Marsden Group objects, which have short orbital periods of around 5.5 to 6.1 years. Peter Jenniskens proposed that the Delta Aquariid stream could have originated from progressive fragmentation of fragments in the Machholz complex, which has evolved over many centuries.

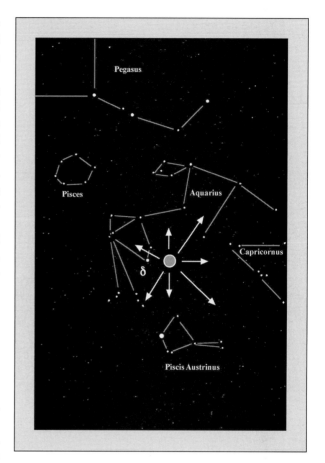

Perseids ($L_S = 140.19°$)

The Perseids are the perfect summer meteor shower and have been enjoyed by countless millions over the centuries. Indeed, they have been recorded as far back as AD 36 when the Chinese chronicled that "more than 100 stars fell."[5] Dante (1265–1321) may have made reference to them in his *Purgatorio* (Canto V):

> Vapours enkindled saw I ne'er so swiftly
> At early nightfall cleave the air serene,
> Nor, at the set of the sun, the clouds of August,

As early as 1836, Belgian astronomer Adolphe Quételet called attention to the fact that every August a large number of meteors come from Camelopardalis. He is not in

[5] From Ichero Hasegawa's (Otemae Junior College, Japan) paper, "Historical Records of Meteor Showers," published in the Proceedings of the International Astronomical Symposium held at Smolenice, Slovakia, July 6–12, 1992, Bratislava: Astromical Institute, Slovak Academy of Sciences, 1993, edited by J. Stohl and I. P. Williams, page 209.

error; over the course of a month, the Perseid radiant drifts through Perseus, Cassiopeia, and Camelopardalis. But perhaps he should more clearly be identified as the first person to note a radiant, because the shower's periodicity had already been known and documented for centuries. In his 1827 *Pocket Encyclopedia of Natural Phenomena*, for instance, Thomas Ignatius Maria Forster includes a "Rustic Calendar" section; under the date of August 10 it reads: "Falling [s]tars and Meteors most abound [around] this time of year."

Forster also informed Quételet that, among the Irish, the meteors were known as the "burning tears of St. Lawrence," whose annual feast is kept on the tenth day of that month. According to legend, on August 10, 258, St. Lawrence was persecuted under the Roman emperor Valerian and executed on a red-hot gridiron.[6] Today, the Perseid maximum falls on August 12–13 UT, still very near that feast.

A delightful midsummer shower, the Perseids remain active from July 17 to August 24 and have many attractive

[6] Bishop St. Ambrose of Milan, the Latin poet Prudentius, and others, tell us that Saint Lawrence refused to give in to his torturers, saying, "I am cooked on that side; turn me over, and eat." (Today, Saint Lawrence is the patron saint of comics, among other things.) Legend aside, many modern scholars favor death by decapitation as the more likely form of the the saint's execution. Also noteworthy is that the remains of the great German-born, English astronomer William Herschel (1738–1822) is interred in the church of St. Lawrence at Upton – beneath an epitaph that reads: *coelorum perrupit claustra* (he broke through the barriers of the heavens).

features: warm temperatures across the Northern Hemisphere make meteor vigils this time of year enjoyable; its radiant, which is very near the famous Double Cluster in Perseus near maximum, is already up when the Sun goes down; the shower's zenithal hourly rates can, on occasion, exceed 100; and the meteors are typically dynamic, being very swift, bright, and colorful; many burn and spit flames on long paths and leave long-lasting trains, which make binocular observing of them highly recommended.

The shower's parent comet, 109P/Swift-Tuttle, orbits the Sun every 130 years or so. "At 26 km (16 miles) across," Jenniskens says, "this is the biggest known comet to come close to Earth's orbit, and 109P/Swift-Tuttle has been doing so for at least 160,000 to Earth's orbit. Hence the very reliable summertime shower. Most meteoroids in the main stream peak around August 12 are about 5,000 years old, while the long July tail of the shower's activity dates much further back in time."

Its last perihelion passage in 1992 brought much excitement. In 1991, the shower produced extremely high zenithal hourly rates of 550 – the highest rates on record. Alas, as you read this, the comet is returning to the outer Solar System, so significant outbursts should now be on the wane. Still, you can expect typical rates of a meteor every minute or so around the time of maximum. Meteor counts drop sharply after that time, with ZHRs of about 10. By the way, the Perseids can also display secondary and tertiary peaks. So observers around the world have a good chance of catching at least one of these maxima.

The Perseids put on a fascinating show on the morning of August 12, 2001, over Hawaii. Activity came in spurts with two meteors coming in quick succession of one another every 10 or so minutes, with more dramatic bursts of four or more meteors flashing nearly simultaneously every 20 to 30 minutes.

The meteors displayed a bell-shaped distribution of magnitudes. Of the 133 meteors observed from 12:00–15:00 UT (with one 15-minute interval of cloud), 72 percent of the meteors blazed between 3rd and 1st magnitude, were swift, and left brief trains. Sixty percent of the meteors were 2nd magnitude and fainter, with half of them shining at magnitude 3. Nearly forty percent of all meteors observed shined between 1st magnitude and magnitude 0.

The two brightest meteors were fantastic spectacles. At 13:39 UT, a −2 green meteor flew to the north of the radiant, exploding several times along the way. Its train was a highly luminous green smoke that began *sparkling* about half a second after the terminal burst. The sparkling lasted for about a second, and it looked like a string of firecrackers going off in the train. I had never witnessed anything like it before.

The brightest meteor was a magnitude −5 fireball that exploded in Cassiopeia about 15° west of the radiant. The flash from the terminal burst was an incredibly bright white "strobe" that simultaneously lit up the landscape.

Through binoculars, the train was equally amazing, for it was bright green, dense, and coiled; it initially rivaled Vega in brilliance (magnitude 0). Amazingly, this train also scintillated or sparkled for about two seconds before fading away. This sparkling was on a finer scale than that of the −2 meteor mentioned above, being more granular while the other was more bombastic. I followed fireballs well into astronomical twilight. Although my watch ended at 15:06 UT, I'm certain I would have seen more had I lingered.

With these brilliant spectacles in mind, do not let the Moon deter you from a Perseid watch. One of my most memorable Perseids occurred over Hawaii on the morning of August 12, 2000. At 3:07 am, I had just stepped outside when I saw a green fireball descend from the zenith toward the west. Flames dripped from its head as it ripped into a 22° halo, which surrounded a pale-orange waxing-gibbous Moon near setting. In the following hour, I tallied 40 meteors, 50 percent of which were brighter than 2nd magnitude; 40 percent of them were greater than 1st. Such activity is typical each year around the time of maximum.

Draconids ($L_S = 196.4°$)

The Draconids are of great interest to meteor observers because the shower, which is normally placid (only a few meteors per hour) periodically bursts to storm level. Twice in the twentieth century (once in 1933, then again in 1946) observers saw spectacular displays with ZHRs exceeding 500. Both events occurred when the Earth passed through the debris stream of its parent comet, 21P/Giacobini–Zinner (which is why the shower is sometimes referred to as the Giacobinids), shortly after the comet passed the junction of the two orbits, which it does every 6.6 years. More importantly, Jupiter had steered the dust trails smack into Earth's path at that time.

The first storm took place on October 9, 1933, when Earth passed the junction of the orbits 80 days after the comet. On that night the ZHRs peaked at 10,000. Observers at the right longitude witnessed one of the twentieth century's few great meteor storms, with more than 150 meteors per minute! Albert Antonie Nijland noted in the 1935 *Bulletin of the Astronomical Institute of the Netherlands* (volume 7, page 248) that "[b]y far the greater part of the meteors near the *Dragon's* head had very short paths, and moved slowly, many of them being even nearly stationary." The others had longer paths and moved faster. The meteors ranged in brightness from 4th magnitude to that of Jupiter. Gary Kronk (http://meteorshowersonline.com) includes a most incredible report from Birchircara, Malta, where R. Forbes-Bentley observed over 22,500 meteors in just a few hours and estimated a peak rate of 480 per minute at 20:15 UT.

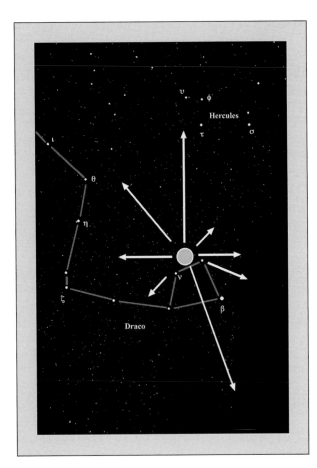

The second storm occurred on October 9–10, 1946, when Earth crossed the comet's orbit only 15 days behind it. Counts were as high as 200 meteors per minute over the Americas. The 1946 storm was also the first definitive meteor shower detected by radar; during the shower's maximum, hundreds of radio "pings" were detected every minute as the meteors streaked through Earth's high atmosphere, ionizing the air around them and leaving behind a trail that can briefly reflect radio waves. (Commonly, radio receivers can detect up to 10 times more meteors than visual observers under dark skies.)

Aside from the high counts, Kronk says that one visual highlight of this event was the appearance of a large blue-white fireball over Southern California, which left a yellow train that lasted over three minutes. "As the train drifted and became diffuse," Kronk says, "it took on the shape of a horseshoe."

Other years with noteworthy activity were 1952 (ZHR = 250), 1985 (ZHR = 550) and 1998 (ZHR = 300). The IMO also notes that a largely unexpected outburst occurred in 2005, probably due to material shed from 21P/Giacobini–Zinner in 1946. Although the visual ZHRs in 2005 were roughly 35, this is a much higher value than in years without storm activity, when ZHRs can be as low as 5.

Storm or not, watching for Draconids/Giacobinids is always a worthwhile pastime, especially for binocular viewers. Draconids are especially slow moving [20.5 km/sec (13 miles /sec)], radiating from a point near Nu Draconis inside the Dragon's trapezoidal head,

making them easier than most to catch with optical assistance. Also, especially during good shows, point meteors and those with short paths may be plentiful near the radiant, which from mid-northern latitudes and higher, is circumpolar.

The shower can be dated to the year AD 585, when the Chinese chronicled that hundreds of meteors fell. Since we do not know the radiant point of the shower, it's also possible that this is a chronicle of Orionids instead (see below).

Orionids ($L_S = 208.6^\circ$)

The Orionid shower is the autumn counterpart to the dependable springtime Aquarid meteor shower. Both showers share the same parent: Comet 1P/Halley. As explained on page 132, Earth intersects Comet Halley's orbit twice each year: once near the comet's descending node in April–May (producing the Eta Aquarid meteor shower), then again near the comet's ascending node in October–November (producing the Orionid meteor shower). The Orionids are active from October 2 to November 7; they peak around October 21.

Like the Eta Aquarids, the Orionids move in retrograde orbits, so they collide nearly head-on with the Earth. Again, these high-energy impacts are responsible not only for the great speed in which they enter the atmosphere [66 km/sec (41 miles/sec)], but also for their exceptionally high percentage (~40%) of long-duration trains; almost all of the meteors brighter than magnitude 0 leave trains; as early as 1877, Denning noted that of the 57 Orionids he recorded from October 16–19, 47 of them left streaks. (All good news for binocular observers interested in these fascinating phenomena!) And while the Orionids have a lower zenithal hourly rate (25–35) than the Eta Aquarids (28–85), the Orionid radiant is much higher in the sky; for Northern Hemisphere observers, that means a greater opportunity for a larger tally.

During four nights in October 2006, the autonomous fireball observatories of the Czech part of the European fireball Network (EN) recorded 48 fireballs belonging to the Orionids. This is significantly more than the total number of Orionids recorded during about a five-decades-long continuous operation of the network. The shower may also have elevated rates every 12 years or so; elevated rates were expected again in 2008. Indeed, conditions were wonderful on the morning of October 21 from my home in Volcano, Hawaii.

The show began immediately with one meteor occurring every minute or more for the first five minutes. In the first hour of observation, beginning at 12:03 UT, I recorded a ZHR of 85.5. Nearly 38 percent of the meteors seen were 1st magnitude or brighter, the brightest being an exploding fireball. The ZHR remained elevated during the second hour, at 67.5, before the dawn. Since the Orionids are also noted for having a broad maximum (with peak activity lasting from October 20–22), I went

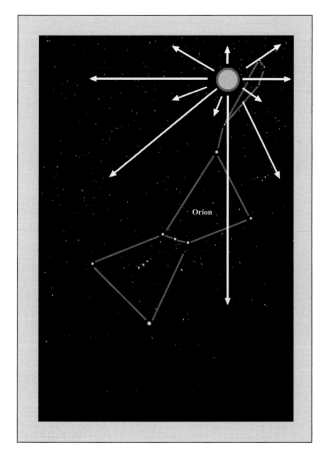

out again the following morning and immediately saw six Orionids in two minutes. The following hour also had a ZHR of 67.5, with 77 percent of the meteors being 3rd magnitude or brighter. The activity came in waves and many of the meteors appeared in pairs. One fireball flashed so intensely that the sky lit up as if by lightning.

Like the Eta Aquarids, the Orionids consist of two components: the one rich in bright meteors that peaked in 2006, and a broader peak rich in fainter meteors that is more stable. As a result, activity sometimes remains roughly constant for several consecutive nights centered on the main broad peak. Look for very swift meteors, many of which are bright, radiating from a broad area of sky in northeastern Orion and southwestern Gemini, between Xi (ψ) Orionis and Gamma (γ) Geminorum. In fact, Denning designated the meteors radiating from this area not only the Orionids, but also the Gamma Geminids.

Leonids ($L_S = 235.1°$)

Arguably the most famous and important meteor shower in history, the Leonids are active from November 14–21, with a maximum occurring around November 17. Although shower rates in some years can be relatively low, with ZHRs achieving 10, the Leonids have long amazed observers with their periodic storms – episodes of which have been seen throughout the ages at roughly 33-year intervals beginning in the year 902, when a shower was seen in Taormina. Throughout the early historical record

it is common to find reports of stars falling "like rain" during storm years.

German naturalist and author Alexander von Humboldt (1769–1859) made the first known detailed report of storm-level activity. On the morning of November 12, 1799, Humboldt and French explorer and botanist Aimé Bonpland (who accompanied Humboldt during five years of travel to the equinoctial regions of the Americas) were on the Cumaná coast of Venezuela, when they witnessed the following spectacle:

> A little before dawn, at about half past two in the morning, extraordinarily luminous meteors were seen. Bonpland, who had got up to get some fresh air in the gallery, was the first to notice them. Thousands of fire-balls and shooting stars fell continually over four hours from north to south. According to Bonpland, from the start of this phenomenon there was not a patch of sky the size of three-quarters of the moon that was not packed with fire-balls and shooting stars. The meteors trailed behind them long luminous traces whose phosphorescence lasted some eight seconds.

Not until Humboldt published this account did the idea that meteors might originate from one location in the sky come to fruition. Still, it took another decade or so before meteor study became a respected discipline.

But the greatest Leonid spectacle was yet to come. On the morning of November 12, 1833, observers across the eastern United States, from the West Indies to Canada saw meteors falling so thickly that some estimated a thousand meteor flashes each minute. In his delightful and informative 1974 book, *Comets, Meteorites & Men* (Taplinger Publishing Co., Inc.; New York), Peter Lancaster Brown includes the alarming experience of a cotton planter in South Carolina:

> I was suddenly awakened by the most distressing cries that ever fell on my ears. Shrieks of horror and cries of mercy I could hear . . . While earnestly listening for the cause, I heard a faint voice near the door calling my name. I arose, and taking my sword, stood at the door. At this moment I heard the same voice beseeching me to rise, and saying, "Oh, my God! the world is on fire!" I then opened the door and it was difficult to say which excited me most – the awefulness of the scene, or the distressed cries . . . Upwards of one hundred lay prostrate on the ground; some speechless, and some uttering the bitterest cries, but most with their hands raised, imploring God to save the world and them. The scene was truly aweful; for never did rain fall much thicker than the meteors fell towards the earth, – east, west, north, and south, it was the same!

A famous depiction of the 1833 Leonid storm opens this chapter. It was produced for the 1889 Adventist book, *Bible Readings for the Home Circle* (Review and Herald Publishing Association, Maryland). Adolf Vollmy created the

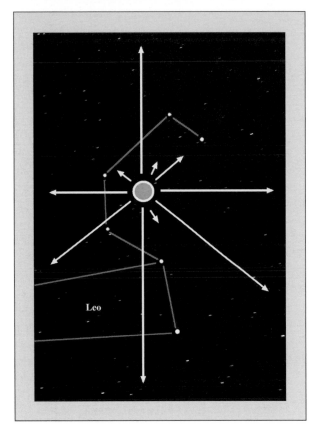

Leo

long time researchers could not decide on what fraction of 33 would correspond to the correct period."

Three decades later, Hubert Anson Newton of Yale College furthered this theory by proposing that only the 33.3-year period could account for the changing time of the peak over the years. He then made the bold prediction that another storm would occur in 1866 . . . and it did. That year, on the night of November 13–14, Robert Ball, an observer at Birr Castle in Ireland, said, "It would be impossible to say how many thousands of meteors were seen, each one of which was bright enough to have elicited a note of admiration on any ordinary night . . ."

While credit for the the discovery of the storm's 33-year periodicity clearly goes to Newton, it's important to consider these displays from the native and ancestral perspective. For instance, Humboldt noted the following about the Cumaná after the 1799 Leonid storm:

> Almost all Cumaná's inhabitants witnessed this phenomenon as they got up before four in the morning to go to first mass. The sight of these fire-balls did not leave them indifferent, far to the contrary; the older ones recalled that the great 1766 earthquake was preceded by a similar manifestation.

While Leonid storms are spectacular, they can be brief. On the morning of November 17, 1966, for instance, a magnificent Leonid storm occurred over North America. Although I did not see it, a friend of mine, the late Dennis Milon, did. He was with 13 other observers at Kitt Peak National Observatory in Arizona when the outburst occurred. He later told me that the meteors were pouring out of the radiant so rapidly and successively that he could sense the Earth moving through space. "It was like being in a car at night while driving through a snowstorm," he said. Making an accurate count was impossible, but Milon estimated that, during peak activity, which lasted about 20 minutes, a single observer would have seen some 2,500 meteors per minute, or about 40 meteors per second. Other observers in equally dark-sky locations

engraving, which is a reproduction of an original painting by Swiss artist Karl Jauslin; Jauslin's painting is based on a verbal account by Joseph Harvey Waggoner – a minister – on his way from Florida to New Orleans.

The following year, American astronomers Denison Olmsted and Alexander Catlin Twining, suggested that the annual Leonid showers were the result of Earth passing through a cloud of meteor particles each November. Humboldt's account was then remembered and also that he had said that a similar storm was seen "thirty years previously on the table-land of the Andes." Jenniskens notes, however, that while a "33-year period was implied, for a

estimated a maximum of only about 1,000 meteors per minute. Regardless, Milon, who published his experience in great detail in a 1967 article in *Journal of the British Astronomical Association* (volume 17, page 89), said that "It was obvious to us that this type of shower would terrify the ignorant . . ."

The Leonids did not achieve visual storm levels in 1999, but the activity that year, and the next few years, was memorable. The following account is from my 2008 companion book, *Observing the Night Sky with Binoculars* (Cambridge University Press; Cambridge):

> The most memorable activity for me occurred in the predawn hours of November 16, 1998 (my birthday). I was not planning to observe that morning. My wife, Donna was away in Boston, and I in bed, asleep, with our late Pomeranian, Pele, curled up on Donna's pillow. But I awakened at 5:00 am to a weird, soundless flashing. When I looked out the bedroom window, the entire yard suddenly lit up as if by lightning. Just as my mind screamed, "What was that!," I saw a brilliant fireball fall from the heavens. Quickly I grabbed Pele, wrapped her up in a blanket and dashed outside to my backyard. In the next 45 minutes, I saw 30 dazzling fireballs and four other meteors raining down from the Sickle of Leo, which was high overhead. With each falling star came a brilliant visual report that bathed the night in an eerie glow (see the photo illustration below). The magic of that morning was in the drama of the unexpected; the meteor activity on the following morning, the morning of the predicted peak, did not compare to that glorious spectacle. It was a valuable reminder that prediction is a theory not a science, and that we as observers should always expect the unexpected.

The Leonids' parent comet is 55P/Tempel–Tuttle (1866 I), and its stream is laced with filaments, which can lead to enhanced displays. The shower's radiant lies in the famous Sickle of Leo, which for observers at mid-northern latitudes rises around midnight. So the first meteors to strike the Earth's upper atmosphere do so at a glancing angle, leading sometimes to extremely dramatic fireballs with extremely long paths. The Leonids have the fastest known velocities and are noted for leaving long-lived, greenish trains. The beauty, number, and longevity of these trains may be unrivaled in binoculars (see page 120).

Geminids ($L_S = 262.08°$)

On its website, the IMO calls the December Geminid shower "One of the finest, and probably the most reliable, of the major annual showers presently observable . . . Even from more southerly sites, this is a splendid shower of often bright, medium-speed meteors, a rewarding sight for all watchers, whatever method they employ." And I have to agree. Moon or not, the Geminids always have

something to offer. Unlike the extreme highs and lows of some of the more dramatic, episodic meteor showers, like the Leonids, the Geminid displays are like a fire in the hearth that just keeps a warm steady glow.

Because Geminid peak ZHRs are very stable at around 120 from year to year, it's common for an individual to see several of its colorful fireballs every hour . . . sometimes more!

In a 1900 issue of *The Observatory* (volume 23, page 366), Walter Ernest Besley of England's Clapham Common, notes that sometime around December 11–13, 1833, "as many as ten meteors were seen simultaneously" – a phenomenon that repeated itself three years later on December 11. And on the evening of December 10–11, 1841, he notes that "Colla, at Parma," saw in one-half hour "23 very brilliant meteors, nearly all of them having luminous trains."

The radiant is easy to find, being near the 1.6-magnitude star Alpha (α) feminorum (Castor) in Gemini the Twins. As seen from mid-northern latitudes, the radiant rises shortly after sunset and is high in the sky in the post-midnight hours before dawn. The shower lasts from December 7 until the 17, with peak activity occurring around December 13–14. Generally, but not always, activity drops off sharply after maximum.

For instance, on December 15, 2003 (UT), I was out observing with my telescope. At about 6:45 UT, I looked up from the eyepiece and saw five Geminids in 30 seconds. While I could not sit back to watch the show (I had other observing plans with my telescope), I did occasionally glance up at the sky throughout the night

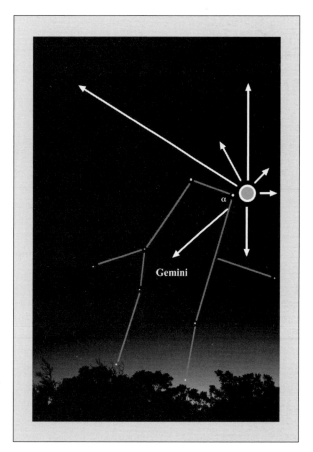

and saw some really nice fireballs until the Moon rose and I left at 9:30 UT. Several times I had to look up from the eyepiece because the ground around me lit up as Geminids burst forth. I also saw several Geminids in the eyepiece while I was making observations of deep-sky objects. The most spectacular Geminid, however, was a magnitude −2 streak that passed directly overhead. That one caused me to leap up from my chair. This meteor had a long and brilliant emerald-green train; it also had a *double* yellow nucleus that tumbled end over end (or one over the other) as it fell.

The IMO notes that the Geminid peak "has shown slight signs of variability in its rates and timing in recent years."

The parent of the Geminid shower remained unknown until 1983, when NASA's Infra-Red Astronomy Satellite (IRAS) found a several-kilometer-wide *asteroid* (3200 Phaethon) moving in an orbit similar to that of the Geminid meteoroids. Indeed, the object was not only confirmed to be the meteor shower's parent but also cataloged as a potentially hazardous Near-Earth Asteroid — one that sails past the Earth at a distance of only eight times that of the Moon. But is Phaeton really an asteroid? Probably not. It's more likely a fragment of an extinct or dormant comet that broke apart sometime in the past. The remnant 3200 Phaethon then went on to parent the Geminids.

The Geminids are a relatively young stream. According to Jenniskens, the stream originated from a breakup of 3200 Phaethon in around AD 1030 and the cloud of meteoroids first intersected the Earth's orbit only at the end of the eighteenth century. The shower is expected to continue increasing in peak rates until about 2050, after which the densest part of the stream moves on.

Artificial satellites

As you look up at the night sky, you might see a "star" moving at a steady and stately pace against the fixed stars. Its motion, you notice, is far too slow to be a meteor, and its course appears too purposeful and long. Indeed, you are most likely sighting an artificial satellite.

Ever since the Soviets launched Sputnik, the world's first artificial satellite, on October 4, 1957, artificial satellites have been orbiting Earth in abundance. According to the the European Space Agency (ESA), as of January 1, 2008, some 6,000 satellites have been placed into orbit since Sputnik. Seventy percent of those cataloged are traveling in low-Earth orbit (LEO), which extends to 2,000 km above the Earth's surface. As of this writing, only about 800 of these satellites are operational. That means that about 90 percent of all satellites launched have ended up as inactive space debris, which continue to orbit the Earth — including fragments of those that have collided, exploded, or otherwise become abandoned.

In fact, as I was writing this paragraph, I received an e-mail from my friend Terry Moseley of the Irish Astronomical Association, alerting me, quite serendipitously, that two major satellites had an unprecedented collision in orbit. On February 3, 2009, a commercial Iridium communications satellite (Iridium 33) and a defunct Russian Cosmos satellite (Cosmos 2251) smashed into each other while orbiting 790 kilometers (491 miles) over northern Siberia, creating a cloud of wreckage that contained many dozens, if not hundreds, of fragments. Moseley expresses his concern:

> The worst aspect of this, is that the problem will grow exponentially! . . . I thought maybe I was being too pessimistic when I gave a lecture to the Cork Astronomy Club about 8 years ago on the topic 'Aliens, Why Aren't They Here?' . . . One reason I advanced for their (apparent!) nonappearance was that they might have gone through the phase we may be entering now: we may make our immediate space environment too risky for future manned space missions! Just think of the future if there's major damage to the ISS; even worse if there's a fatality. I suppose if we wait a few thousand years or so, most of the rubbish will have decayed into the atmosphere, but will we have lost the impetus for space exploration by then? I hope not, but it's a risk.

The ESA estimates that more than 12,000 pieces of space junk are orbiting around Earth. At least 11,500 of those (mostly commercial, military, scientific, and navigational satellites) are in LEO, at altitudes ranging from 800 to 1,500 kilometers above Earth's surface. Over the course of decades, their orbits will decay and the craft slip closer to Earth, until, eventually, they burn up in the Earth's atmosphere. So some of the burning objects you see as meteors and fireballs may actually be orbital debris re-entering Earth's atmosphere in a fiery death plunge. Rarely, space debris falls to Earth. On March 23, 2001, for instance, the remains of the Russian Space Station Mir met a blazing and noisy end as it thundered into Earth's atmosphere before it showered the southern Pacific Ocean with debris estimated to weigh as much as 25 tonnes!

The ESA image below is an artist's impression of the space junk situation based on actual data. The image shows the debris at an exaggerated size to make them visible at the scale shown. The ESA estimates that the amount

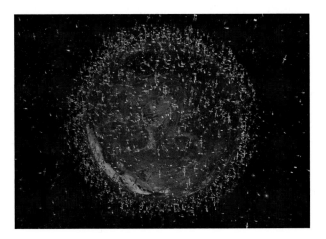

of space junk is increasing by at least 200 per year. Aside from the collisions and explosions, space debris can also be the product of lost or discarded material from space flights and rockets. For instance, on November 18, 2008, Space Shuttle Endeavor astronaut Heide Stefanyshyn-Piper lost her grip on a tool bag during a space walk outside the International Space Station (ISS). The shiny bag floated away, becoming an official piece of space junk. Sunlight reflecting off its highly reflective surface made it visible to skywatchers across North America and Europe. Its brightness, at times, achieved magnitude 6.4, making it visible to keen naked-eye observers under extremely dark skies and binocular users. Amateur astronomer Edward Light saw it through 10 × 50 binoculars from his backyard in Lakewood, New Jersey.

Like the tool bag, we see satellites because sunlight reflects off their shiny surfaces. The range in brightness is great and varies depending on a number of factors, including the size of the craft or debris, its distance from Earth' surface, the reflectivity of its parts, and the angle at which we see it. Some satellites, like Iridium communication satellites, normally sail across the sky at 6th magnitude (a good binocular target), but they can also suddenly flare up to magnitude −8, which is 30 times brighter than Venus at maximum magnitude. After some 5 to 20 seconds or so, the satellite fades back to its normal brightness.

Iridium flares are astonishing and can even be detected in daylight, especially around the time of sunrise or sunset. The flares occur whenever sunlight glints off of an Iridium satellite's long and flat aluminum reflectors, of which there are three. To find out more about all these satellites, when to see them, and how to contribute observations, check the Visual Satellite Observer Home Page (http://www.satobs.org/iridium.html), which also provides links to other artificial-satellite sites.

The ISS is the biggest and brightest artificial craft orbiting the Earth. The football-field-sized station is comprised of several metallic body parts and large solar panels. Depending on the angle of illumination, it can shine as brightly as magnitude −3 (about as bright as the planet Jupiter) or as dim as magnitude +1.5 (the magnitude of the star Castor in Gemini the Twins or the planet Mars, when dim). Though smaller, NASA's Space Shuttles can be nearly as impressive as the ISS when they sail overhead, but nothing compares to seeing the Space Shuttle attached to the ISS; the combined light of which can rival or exceed Venus in brightness under perfect viewing conditions. These crafts (individually or together) are easily spied with the naked eye and binoculars, appearing as a bright star that doesn't twinkle, moving gracefully and silently across the sky like something sliding gently across smooth black ice.

Now for an interesting series of "ifs": if you own a pair of image-stabilized binoculars that magnify 10× or more; or if you are positioned so that you are comfortable (say lying back on a recliner) and can steady the binoculars; and if the ISS passes overhead in the twilight (which will

cut down on glare); you may be able to discern a shape to the craft, and perhaps resolve its solar panels! What I'd like to know, however, is if you can see through your binoculars an astronaut performing a spacewalk outside the ISS or Space Shuttle? If a tool bag can be seen in binoculars, why not a man, or woman, in a space suit?

Unlike with meteors, which flash and vanish in a matter of seconds, satellites can take minutes to cross the sky. Generally, they are best seen shortly before or after astronomical twilight, or about 45 minutes before or after sunrise. They can be seen throughout the night, though only the ones in the highest of orbits are visible late at night. Again, most satellites are in LEO, and these are best seen in the early evening or early morning hours. The long-exposure image below shows the trail of the International Space Station in the early evening as it moved across the sky from the south (left) to the northwest.

It's common to see satellites behave "mysteriously." In fact, misinterpretations of their "odd" behaviors have led some to believe that they've witnessed a craft from outer space. For instance. It's common to see a satellite suddenly fade and vanish right before your eyes. To the uninitiated, it looks as if a visiting UFO suddenly took off into hyperspace − rapidly receding from view (the dimming light) until it's gone! Actually, what happened is that the satellite entered the Earth's shadow and was eclipsed. The dimming is the satellite passing through the graded penumbral shadow, before it is totally eclipsed (and blinks out) in the umbral shadow.

Many satellites rotate. As they do so, different parts of their surface can reflect light, causing you to see a satellite blinking at regular intervals as it crosses the sky – as if it's generating some alien morse code. Some satellites can leak fuel, or be hit by micrometeoroids, which can suddenly affect the rotation of the satellite and cause it to tumble, thereby creating a very erratic flash pattern as it crosses the sky.

There is so much to see and experience in the sky, all of it exciting. If you want to know when and where to look for the ISS, Hubble Space Telescope, or other bright orbiting satellites, there are a number of websites dedicated to the search. One of the most popular is Heavens Above (http://www.heavens-above.com/), which requires that you know your local coordinates. You could also use Spaceweather.com's Satellite Tracker website (http://www.spaceweather.com/flybys/?PHPSESSID=u2pkarift2oeg8780ug0a9d0e4); This program will give you not only the times of passage for the ISS and Space Shuttle but also a listing of a half-dozen or so of the most interesting satellites visible from your location including the ISS tool bag and Omid (Iran's first satellite). All you have to do is enter your zip code.

Of course, for ISS and Space Shuttle visibility predictions, you could also go to NASA's dedicated website (http://spaceflight.nasa.gov/realdata/sightings/); all you need to do is select the country you're in, select the town, and *voila*, you get a list of which craft is visible, how long it will be visible, its maximum elevation, and the altitude and azimuth of its approach and departure.

Altitude is an object's height above the horizon: An object on the horizon has an altitude of 0°; an object directly overhead (*zenith*) has an altitude of 90°. *Azimuth* is a measure of the object's compass direction along the horizon as measured clockwise from north: an object due north has an azimuth of 0°/360°; due east (90°); due south (180°); due west (270°). So if the ISS was to appear at an altitude of 45° and an azimuth of 90°, you would look for a bright moving star half way up the eastern sky around the predicted time.

Some sources will give you the object's magnitude. Roughly, 6th magnitude is the faintest star one can casually see under a very dark sky. The stars of the Big Dipper, or Plough, shine around 2nd magnitude. And Venus, the brightest celestial object in the sky other than the Sun or Moon, shines at magnitude −4.5 when brightest. Happy hunting!

The 100+ brightest objects in the Main Asteroid Belt

Below are the brightness ranges and mean opposition magnitudes for the first 100 numbered objects in the Main Asteroid Belt, along with a couple of bright, higher-numbered objects. They are listed in order of maximum brightness. For objects on nearly circular orbits, the mean opposition brightness will be (nearly) the average of the object's maximum brightness and its minimum brightness. For more eccentric orbits, the mean opposition magnitude will be skewed towards the object's minimum brightness.

Handheld binoculars refer to those in the 7×35 to 10×50 range. Medium-sized, mounted binoculars refer to those in the 12×50 to 16×70 range. Large binoculars range from 20×80 to 25×100. Those using handheld binoculars under dark skies (with a naked-eye magnitude limit of magnitude 6.0 or better) should also try for some of the brighter targets in the medium-sized binocular table. Likewise, those using medium-sized binoculars under dark skies should try for some of the brighter objects in the large-binocular table. The magnitude limit you achieve on any given night will depend on sky conditions and other factors, such as sky transparency, moonlight, light pollution, optical quality, state of health, and observing experience.

Best targets for handheld binoculars (when the object is at, or near, maximum brightness)

Object	Brightness range (Max.)–(Min.)	Mean brightness (at opposition)
4 Vesta	5.3–6.5	6.0
1 Ceres (dwarf planet)	6.7–7.7	7.2
7 Iris	6.7–9.5	8.5
2 Pallas	6.7–9.7	8.6
3 Juno	7.5–10.2	9.1
6 Hebe	7.7–10.0	9.1
18 Melpomene	7.7–10.4	9.4
15 Eunomia	7.9–9.9	8.9
8 Flora	8.0–9.8	9.0
324 Bamberga	8.1–12.1	10.9
19 Metis	8.2–9.7	9.2
192 Nausikaa	8.2–11.3	10.3
20 Massalia	8.4–10.1	9.4
27 Eurtepe	8.4–10.6	9.7
12 Victoria	8.6–11.2	10.2
29 Amphitrite	8.7–9.6	9.2
11 Parthenope	8.8–10.1	9.6
5 Astraea	8.8–11.1	10.3
43 Ariadne	8.8–11.1	10.3
89 Julia	8.8–11.2	10.3
39 Laetitia	8.9–10.4	9.8
44 Nysa	8.9–10.7	10.0
19 Fortuna	8.9–10.9	10.2
10 Hygiea	9.0–10.3	9.8
14 Irene	9.0–10.7	9.9

Good targets for medium-sized, mounted binoculars (when the object is at, or near, maximum brightness)

Object	Brightness range (Max.)–(Min.)	Mean brightness (at opposition)
23 Thalia	9.1–11.7	10.7
42 Isis	9.1–11.8	10.7
40 Harmonia	9.3–9.9	9.6
16 Psyche	9.3–10.6	10.0
21 Lutetia	9.3–11.2	10.4
41 Daphne	9.3–2.5	11.5
80 Sappho	9.4–11.9	11.0
68 Hesperia	9.5–11.6	10.7
51 Virginia	9.6–10.7	10.4
30 Urania	9.6–11.0	10.4
63 Ausonia	9.6–11.1	10.4
511 Davida	9.6–11.8	11.0
79 Eurynome	9.6–12.0	11.2
28 Bellona	9.7–11.6	11.0
37 Fides	9.7–11.6	10.8
88 Thisbe	9.7–11.6	10.9
13 Egeria	9.8–11.0	10.4
22 Calliope	9.8–11.0	10.6
52 Nemausa	9.9–11.2	10.7
704 Interamnia	9.9–11.4	10.6
17 Thetis	9.9–11.5	10.8
33 Polyhymnia	9.9–14.0	12.7
54 Calypso	10.0–12.4	11.6
25 Phocaea	10.0–12.5	11.3
97 Clotho	10.0–12.7	11.6
84 Clio	10.1–13.2	12.3
46 Hestia	10.2–12.2	11.5
85 Io	10.2–12.3	11.4
75 Eurydice	10.2–14.0	12.8
67 Asia	10.2–12.3	11.4
31 Euphrosyne	10.2–12.7	11.6
26 Proserpine	10.3–11.4	10.9
32 Pomona	10.3–11.4	10.9
55 Alexandria	10.3–12.0	11.5
60 Echo	10.3–12.4	11.5
78 Diana	10.3–12.8	11.9
69 Leto	10.4–2.1	11.3
71 Feronia	10.4–12.2	11.3
82 Alcmene	10.4–12.9	11.9
50 Pales	10.4–13.8	12.7
36 Atalanta	10.4–13.9	12.7
61 Danae	10.5–12.6	12.0
74 Galatea	10.5–13.4	12.4

Good objects for large, mounted binoculars (when the object is at, or near, maximum brightness)

Object	Brightness range (Max.)–(Min.)	Mean brightness (at opposition)
64 Angelina	10.6–12.0	11.4
24 Themis	10.6–12.1	11.5
57 Mnemosyne	10.6–12.2	11.6
70 Panopea	10.6–12.5	11.6
49 Doris	10.6–13.1	12.1
56 Pandora	10.6–13.2	12.2
45 Eugenia	10.7–11.6	11.1
92 Undina	10.7–11.9	11.3
65 Cybele	10.7–12.2	11.5
59 Elpis	10.7–12.3	11.7
100 Hekate	10.7–12.6	12.0
93 Minerva	10.8–12.3	11.6
72 Niobe	10.9–12.3	11.6
96 Aegle	10.9–12.7	12.0
38 Leda	10.9–12.8	12.1
53 Europa	10.9–13.2	12.3
83 Beatrix	11.0–12.1	11.7
98 Ianthe	11.0–13.4	12.6
48 Aglaia	11.1–11.9	11.4
47 Melete	11.1–12.6	11.9
81 Terpsichore	11.1–13.4	12.5
77 Frigga	11.2–12.7	12.1
35 Leucothea	11.2–13.7	12.8
90 Antiope	11.2–13.0	12.3
87 Sylvia	11.3–12.2	11.7
34 Circe	11.3–12.6	12.1
95 Arethusa	11.3–13.1	12.4
91 Aegina	11.4–12.6	12.1
94 Aurora	11.4–12.6	12.1
76 Freia	11.5–13.6	12.7
66 Maia	11.6–13.5	12.8
86 Semele	11.6–13.9	13.0
99 Dike	11.7–14.1	13.4
73 Clytie	12.0–12.6	12.3
58 Concordia	12.0–12.7	12.5
62 Erato	12.0–13.9	13.1

Photo credits

Chapter 1

Stephen James O'Meara
Stephen James O'Meara
Stephen James O'Meara
Stephen James O'Meara
Stephen James O'Meara
Donna O'Meara
Stephen James O'Meara
Stephen James O'Meara
Stephen James O'Meara
SOHO (ESA & NASA)
SOHO (ESA & NASA)
Royal Swedish Academy of Sciences / Oddbjorn
   Engvold, Jun Elin Wiik, Luc Rouppe van der Voort,
   Oslo
Stephen James O'Meara
SOHO (ESA & NASA)
Stephen James O'Meara/(facsimile of Harriot's
   observations reproduced in a 1916 *Popular Astronomy*
   article titled "The History of the Discovery of the
   Solar Spots," by Walter M. Mitchell)
Royal Swedish Academy of Sciences / Göran B.
   Scharmer, Boris V. Gudiksen, Dan Kiselman,
   Mats G. Löfdahl, and Luc H. M. Rouppe van der
   Voort.
Stephen James O'Meara
SOHO (ESA & NASA)
SOHO (ESA & NASA)
Stephen James O'Meara (from Maunder's paper
   "Magnetic declination, secular change in, and
   suggested connection with sun-spot activity"
   published in *Monthly Notices of the Royal Astronomical Society*,
   1904)
SOHO (ESA & NASA)
SOHO (ESA & NASA)
NASA
Stephen James O'Meara (from *The Story of the Solar System*,
   D. Appleton and Company, New York, 1908; George
   F. Chambers)
SOHO (ESA & NASA)
SOHO (ESA & NASA)

Chapter 2

Stephen James O'Meara
Stephen James O'Meara
Stephen James O'Meara
Stephen James O'Meara

Stephen James O'Meara
Stephen James O'Meara
Stephen James O'Meara
Stephen James O'Meara
NASA
NASA
Stephen James O'Meara
Stephen James O'Meara
Stephen James O'Meara
Stephen James O'Meara
Stephen James O'Meara
Stephen James O'Meara
Stephen James O'Meara
Stephen James O'Meara
Stephen James O'Meara
Stephen James O'Meara
Stephen James O'Meara

Chapter 3

Both by Stephen James O'Meara
Stephen James O'Meara
Stephen James O'Meara (from *Wonders of the Earth, Sea and
   Sky*, Hall & Locke Company, Boston, 1902; edited by
   E. S. Holden); the illustration accompanies a chapter
   in the book by E. S. Holden (the total Solar eclipse of
   1883), which first appeared in the *Atlantic Monthly*,
   May, 1890
All Stephen James O'Meara (photos and photo
   illustrations)
NASA (Earth); Stephen James O'Meara (Moon)
Both Stephen James O'Meara
All Stephen James O'Meara
Stephen James O'Meara
Stephen James O'Meara
Stephen James O'Meara (photo illustration)
Stephen James O'Meara (photo illustration)
Stephen James O'Meara
Stephen James O'Meara/ Stephen James O'Meara (from
   Astronomy by Observation, D. Appleton and Company,
   1886; Eliza Brown; image of 1867 corona as observed
   by Grosch)
Both NASA
Stephen James O'Meara
Stephen James O'Meara (photo illustration)
Stephen James O'Meara (photo illustration)
All Stephen James O'Meara
Stephen James O'Meara

Stephen James O'Meara
Stephen James O'Meara (photo illustration)
Stephen James O'Meara

Chapter 4
NASA/JPL/Space Science Institute
Stephen James O'Meara (private collection)
NASA
All Stephen James O'Meara
NASA
Courtesy NASA/JPL-Caltech
Stephen James O'Meara
Stephen James O'Meara
Stephen James O'Meara
All Stephen James O'Meara
NASA/JPL
NASA/JPL
NASA/JPL/USGS
Both by Stephen James O'Meara
Stephen James O'Meara
Stephen James O'Meara
Stephen James O'Meara
Stephen James O'Meara
Stephen James O'Meara (photo illustration)
Stephen James O'Meara (photo illustration)
NASA, ESA, the Hubble Heritage Team (STScI/AURA),
    J. Bell (Cornell University), and M. Wolff (Space
    Science Institute, Boulder)
Both Stephen James O'Meara
Stephen James O'Meara (private collection)
Stephen James O'Meara (photo illustration)
Stephen James O'Meara
Stephen James O'Meara
NASA/JPL
NASA/JPL/Malin Space Science Systems
NASA/Johns Hopkins University Applied Physics
    Laboratory/Southwest Research Institute
Stephen James O'Meara
Stephen James O'Meara
Stephen James O'Meara
NASA/JPL/Space Science Institute
Image by Marty Peterson, based on a 1984 image by
    William K. Hartmann. NASA/JPL/University of
    Colorado
NASA, ESA, and The Hubble Heritage Team
    (STScI/AURA)
Stephen James O'Meara
NASA/JPL/Space Science Institute
Stephen James O'Meara
NASA/JPL
NASA/JPL
Stephen James O'Meara
Stephen James O'Meara
Stephen James O'Meara
Ceres: NASA, ESA, and J. Parker (Southwest Research
    Institute)
Vesta: NASA, ESA, and L. McFadden (University of
    Maryland)

Stephen James O'Meara (photo illustration)
Stephen James O'Meara (photo illustration)

Chapter 5
Stephen James O'Meara (from *Wonders of Earth, Sea and Sky*,
    Hall & Locke Company, Boston, 1902; edited by
    E. S. Holden)
NASA/JPL (from Comets in Ancient Culture at
    http://www.nasa.gov/mission_pages/deepimpact/
    media_ancient.html)
NASA/JPL (from Comets in Ancient Culture at
    http://www.nasa.gov/mission_pages/deepimpact/
    media_ancient.html)
NASA/JPL (from Comets in Ancient Culture at
    http://www.nasa.gov/mission_pages/deepimpact/
    media_ancient.html)
NASA/JPL (from Comets in Ancient Culture at
    http://www.nasa.gov/mission_pages/deepimpact/
    media_ancient.html)
Stephen James O'Meara (photo illustration)
Stephen James O'Meara (from *Astronomy by Observation*,
    D. Appleton and Company, New York, 1886; Eliza
    Brown)
Stephen James O'Meara
Stephen James O'Meara
NASA/JPL-Caltech/UMD
NASA/JPL-Caltech
Stephen James O'Meara
Stephen James O'Meara (photo illustration)
Stephen James O'Meara (photo illustration)
Stephen James O'Meara (photo illustration)
Stephen James O'Meara (photo illustration)
Stephen James O'Meara (photo illustration)
Stephen James O'Meara (photo illustration)
Stephen James O'Meara
Stephen James O'Meara (from *The World of Comets*, Searle
    & Rivington, London, 1877; Amedee Guillemin)
Stephen James O'Meara
NASA, ESA, H. Weaver (JHU/APL), M. Mutchler and Z.
    Levay (STScI)

Chapter 6
Stephen James O'Meara (from *Bible Readings from the Home
    Circle*, Review and Herald Publishing Association,
    Maryland, 1889)
Stephen James O'Meara
Stephen James O'Meara (photo illustration)
Stephen James O'Meara (five meteor train photo
    illustrations)
Stephen James O'Meara (from a 1907 *Astrophysical Journal*
    article "Physical Nature of Meteor Trains" by C. C.
    Trowbridge; reproductions of drawings by W. S.
    Gilman Jr., in 1868)
Stephen James O'Meara (from a 1907 *Astrophysical Journal*
    article "Physical Nature of Meteor Trains" by C. C.
    Trowbridge; reproductions of drawings by W. S.
    Gilman Jr., in 1868)
NASA

All Stephen James O'Meara (photo illustration)

NASA (Earth background image)

NASA (Earth background image)/Stephen James
    O'Meara (photo illustration)/Stephen James O'Meara

Stephen James O'Meara (photo illustration)

Stephen James O'Meara (photo illustration)

Stephen James O'Meara (photo of Burrit's 1835 Atlas,
    owned by author)

Stephen James O'Meara (photo illustration)

Stephen James O'Meara (photo illustration)

Stephen James O'Meara (photo illustration)

Stephen James O'Meara (photo illustration)

Stephen James O'Meara (photo illustration)

Stephen James O'Meara (photo illustration)

Stephen James O'Meara (photo illustration)

Stephen James O'Meara (photo illustration)

Stephen James O'Meara (photo illustration)

Stephen James O'Meara (photo illustration)

Stephen James O'Meara (photo illustration)

Stephen James O'Meara (photo illustration)

Stephen James O'Meara (photo illustration)

ESA

Stephen James O'Meara

Index